AN INTRODU
SATELLITE COM

Other Titles of Interest

BP293 An Introduction to Radio Wave Propagation

BP312 An Introduction to Microwaves

BP315 An Introduction to the Electromagnetic Wave

AN INTRODUCTION TO SATELLITE COMMUNICATIONS

by
F. A. WILSON
C.G.I.A., C.Eng., F.I.E.E., F.I.M.

BERNARD BABANI (publishing) LTD
THE GRAMPIANS
SHEPHERDS BUSH ROAD
LONDON W6 7NF
ENGLAND

Please Note

Although every care has been taken with the production of this book to ensure that any projects, designs, modifications and/or programs etc. contained herewith, operate in a correct and safe manner and also that any components specified are normally available in Great Britain, the Publishers do not accept responsibility in any way for the failure, including fault in design, of any project, design, modification or program to work correctly or to cause damage to any other equipment that it may be connected to or used in conjunction with, or in respect of any other damage or injury that may be so caused, nor do the Publishers accept responsibility in any way for the failure to obtain specified components.

Notice is also given that if equipment that is still under warranty is modified in any way or used or connected with home-built equipment then that warranty may be void.

First Published – December 1993

British Library Cataloguing in Publication Data
Wilson, F. A.
Introduction to Satellite Communications
I. Title
621.382

ISBN 0 85934 326 X

Printed and bound in Great Britain by Cox & Wyman Ltd, Reading

Preface

Through pathless realms of space
Roll on.

Sir W. S. Gilbert

In the beginning, or perhaps just afterwards, there were satellites, simply heavenly bodies revolving around planets, in fact the word arises from the Latin meaning *an attendant.* The earth then had only one satellite of its own, the moon, while the earth itself was a satellite of the sun. Happily both moon and sun are still around and those of us who like to keep warm and look around are especially indebted to the sun. However recently along came the *artificial* satellites, at first one by one but now there are so many that the description "artificial" has generally been discarded. Although more than a little expensive, their number is increasing relentlessly. They are certainly here (or rather, there) to stay for despite their huge cost, the benefits they bestow are out of this world, in more senses than one. Not only do satellites bring us a host of additional television channels but they tell us what our weather looks like from up above, relay news within seconds, provide telephony for people on the move, solve many of our navigation problems and help our armed services, altogether a fascinating list of achievements.

It is first necessary however to explain why this book has been written when "An Introduction to Satellite Television" (BP195) was published only a few years ago. At that time the Marcopolo satellite for television was being constructed and in the UK we were getting ready to enjoy satellite television based on an advanced system and requiring a dish antenna of only about 35cm diameter. However a loophole in the directive from on high and a certain amount of financial manipulation triumphed so that although the satellite itself eventually got off the ground, the television network it should have provided did not. We now have instead a profusion of television programmes beamed down from the sky from other less powerful satellites requiring dishes of at least 60 cm diameter, three times the size. It might even be suggested that they are also at least three times more unsightly for we can hardly class the parabolic dish as a thing of beauty.

It is therefore hardly right to continue to publish a book extolling the virtues of a new system which as far as the UK is concerned, does not exist. A description has not been entirely omitted however for the system is in use elsewhere in Europe, for that reason we briefly outline the principles here. Also in this new updated edition the opportunity has been taken to widen the scope to include many of the other aspects of satellite transmission.

The techniques involved are fascinating indeed but let us be under no illusions, they are complicated and are undergoing unprecedented development. Yet without burdening ourselves too much with the electronic technology involved, it is possible to gain an all-round awareness of the whole system. After all, in a restaurant it is hardly necessary to understand the chemistry of cooking to be able to enjoy the food. We can then undertake installation of our own home equipment or certainly know what to expect from a dealer who does it for us. The full story involves a multitude of mechanical and electronic disciplines as a glance at the Table of Contents shows. The discussion therefore covers a wide range. Striking a balance between including just sufficient for the *modus operandi* to be understood and on the other hand saturating the reader with detail is not easy. Let us hope that we have got it right. However the many engineers and scientists who require more than just an insight into the whole affair are not forgotten. They delight in what mathematical formulae have to tell so for these dedicated people the related formulae are included in the Appendices. In this way the book attempts to cater for all tastes and all that is required of the reader is that the grey matter is in good working order.

Some updating may be appropriate on the struggle for supremacy between the American *antenna* and the English *aerial.* The English are losing the battle so there are now fewer aerials and more antennas. One would expect the plural to be antennae but no, these only refer to our creepy-crawly friends. To add to our confusion, satellite television has its *dishes.* A dish is an antenna in a more colloquial form so antenna or dish is used as seems appropriate. The two words may seem to be synonymous but although a dish is always an antenna, an antenna is not necessarily a dish.

Satellite communication is new, much of it very new. Accordingly to be up to date we avoid the partially defunct Imperial System (pints, pounds, inches, miles) and hitch our wagon to the metric.

As in all professions some jargon is inevitable so particularly useful is Appendix 1 which contains a glossary of terms.

F. A. Wilson
C.G.I.A., C.Eng., F.I.E.E., F.I.M.

Contents

Chapter 1
SETTING THE SCENE

Whereas not so many years ago the term "satellite" would hardly ever have entered everyday conversation, today it is commonplace. Apart from their burst into the domestic scene to provide us with a multitude of additional television channels, they have many other functions, especially in world-wide communications, meteorology, space research and guiding ships and aircraft. There are in addition many observation or spy satellites flying round but discussion of these is not encouraged. Whoever can have thought just a few decades ago that we, the ordinary people of this world would be able to look up at the clouds and then through the television report look down on them from up above? Such is progress. Truly enormous distances are involved, how lucky we are that the electromagnetic wave on which the whole system depends travels fast – very fast indeed.

1.1 An Overview

The trend in motor car design is now to shapes which reduce friction of the air, such friction on satellites can be ignored for there is very little air or none at all up there. Thus we accept any peculiar shape with no qualms about it being unconventional. In Chapter 3 we will find that there is no such thing as a stationary satellite, all must be on the move otherwise the earth's gravity will cause them to fall. The simple equations of satellite motion follow in the same chapter.

Inside the satellite is an expensive conglomeration of equipment, expensive because it must be of such high quality that failure is unlikely. Antennas abound on the outside both for receiving control and work signals, also for transmitting back to earth. Some satellites merely send information such as weather data but most relay communication signals such as speech, data and television. These receive radio signals from a ground station then transmit them back to a different area on earth.

The question is bound to arise as to why satellites have developed so explosively. Obviously there is a need for research, weather, navigation and even spy satellites but we already have world-wide communication and television abounds everywhere. There are many reasons but perhaps two of the most important arise from the fact firstly that the earth is a sphere (or nearly) and unfortunately radio waves generally prefer to move in straight lines, thus we cannot transmit waves over great distances as indicated in Figure 1.1(i).

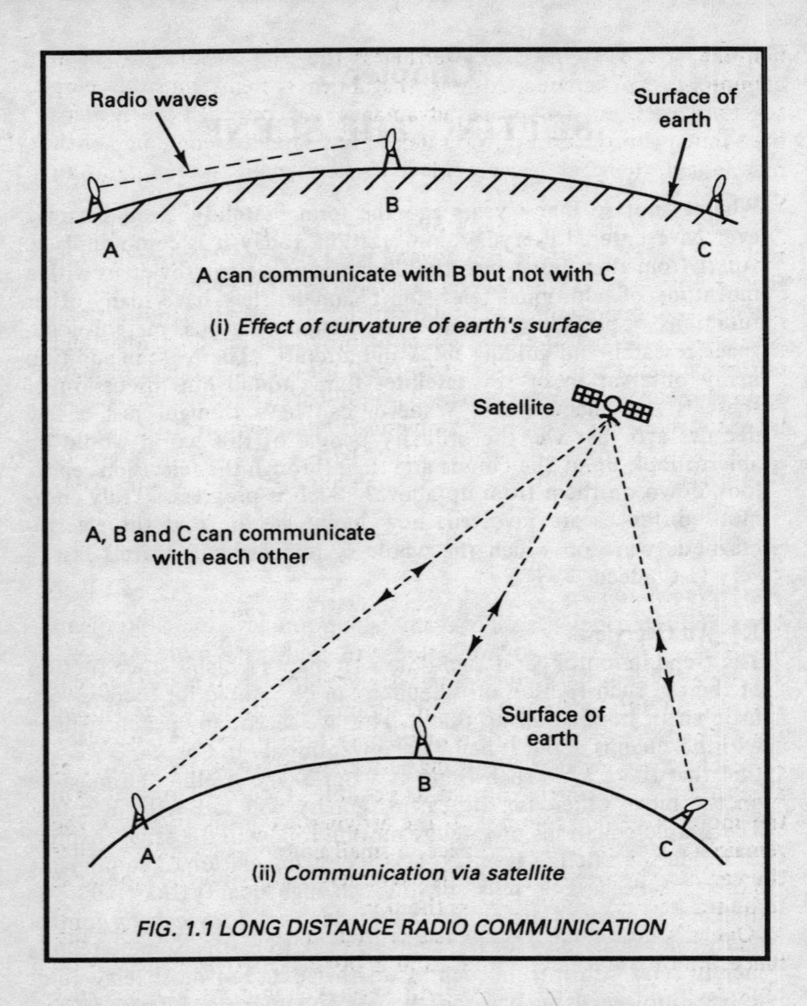

FIG. 1.1 LONG DISTANCE RADIO COMMUNICATION

Alternatively we can use the special mirror-like reflecting properties of the ionosphere, many kilometres up above. Unfortunately the ionosphere is fickle and therefore to a certain extent unreliable. On the other hand a satellite is not so affected and (ii) in the figure shows how much greater distances around the earth can be covered in a single hop by one satellite.

There is another alternative for long distance transmission however, which is the under-sea cable. Circuits are rapidly becoming cheaper now that optical fibre transmission has arrived. In this system thousands of telephone circuits can be transmitted over a few hair-like strands of glass. Nevertheless the irrepressible growth in communication circuits ensures that both systems must develop together, each has its special advantages, e.g. optical fibre systems have much shorter transmission delays but satellites come into their own when there are geographical barriers such as mountains or deserts.

Secondly, the rapid growth of satellite television arises from the fact that a single satellite can cover a large area such as the whole of the UK and Ireland with several programmes. The enormous cost of placing and maintaining a satellite in position is then quite small compared with the alternative cost of providing hundreds of transmitting stations on the ground (over 600 for the UK), this seemingly large number is required because a ground-based signal soon gets lost by absorption, hence the reception area is relatively small.

Overall therefore artificial satellites have a lot to offer, proof of this is the fact that there are several hundred of them up there already!

1.1.1 Moonbounce

Long before artificial satellites came along to alleviate the problems of long distance communication, there was "moonbounce", developed mainly by radio enthusiasts. Figure 1.1(ii) applies except that the satellite is replaced by the moon, used simply for reflecting signals back to earth (and unfortunately also to everywhere else). A highly directional transmitting antenna was required so that as much of the transmitter output as possible reached the moon. On the moon a certain amount of the arriving energy is absorbed, the remainder is scattered into space, a small amount of which reaches the earth. The received signal is extremely weak and the moon has to be tracked as it moves across the sky.

One advantage the system has is that at such an enormous distance, moonbounce communication is possible between stations on opposite sides of the earth. All very well but such a distance involves an insurmountable problem for two-way conversation. Even at the speed of electromagnetic waves (3×10^8 metres per second – a speed with which it is difficult for the imagination to cope), a signal arriving back on earth from the moon has taken over 2.5 seconds for the total journey. This propagation time is doubled between speaking and receipt of an answer, a total time for which conversation is impossible unless very disciplined. Thus moonbounce was not a practical system but it did highlight many of the problems to come as satellite transmission developed.

1.1.2 Early Experimental Satellites

Inevitably moonbounce gave way to a host of experimental satellites put up mainly by the Soviet Union and the U.S.A. One of these was the *passive* (no amplifiers on board) ECHO. This was simply a large balloon with an aluminium coating for reflecting waves back to earth as did the moon but much more efficiently. ECHO roamed the space immediately above the earth (some 1500 km high) and was more efficient as a reflector compared with the moon but tracking it with huge antennas was certainly not easy. Concurrently active satellites were being developed, an early one was COURIER which received and stored teletype messages for re-transmission as required. Then in 1962 the first active satellite capable of immediate re-transmission went up, known as TELSTAR. Signals were transmitted to it in the 6 GHz band and sent back to earth in the 4 GHz band. It had a medium altitude elliptical orbit hence still requiring antennas to follow it. More of this type went up until eventually the breakthrough came when the geostationary (*geo* = earth) orbit was achieved by a SYNCOM satellite. Briefly, although far away, the orbit is such that the satellite circles round above the equator in the same time as the earth rotates once, hence the satellite remains in the same position in the sky *with respect to* earth. The expensive and difficult technique involving antennas to track a satellite moving with respect to earth is therefore not required.

The Syncom satellites were able to prove what could be done with the geostationary system, this led to the launch of the first of the many of the INTELSAT range in 1965, known originally as EARLY BIRD but later renamed INTELSAT 1. Concurrently Soviet Union MOLNIYA satellites were being launched but not then into geostationary orbits, instead into special elliptical orbits more suitable for the enormous expanses of Soviet territory.

Since those days satellites have been going up into orbit regularly, they are able to provide a full range of communication services including television distribution not only to densely populated areas but also to smaller countries and isolated communities.

1.2 Satellite Systems

We now move on to present day systems which have developed from the earlier experimental work. Passive satellites need not be considered further, being mainly relegated to the experimental stages. Having no amplification on board they require very powerful ground transmitters and even then return an extremely weak signal. Active satellites on the other hand have proved their worth. This is clearly evident from the fact that there are so many up there in the geostationary orbit that whereas a few years ago they were spaced 5 degrees apart, this has had to be reduced to 2 degrees to satisfy the increasing demand for new orbital positions.

Generally satellite systems can be subdivided according to the facilities they provide. Most of us naturally think in terms of television programmes. These can only be provided by the geostationary system for the alternative which is to track a satellite moving relative to earth would be unthinkable. Equally important is the use of satellites for communication links for international telephone circuits, also with data transmission for business. In fact, given sufficient traffic requirements, large local business communities may even have their own earth terminals or *teleports*. The wonders of news gathering and distribution by satellite are already with us and again such services are most likely to be provided via the geostationary system.

Mobile systems for ships and aircraft have developed apace, mainly because of the unreliable nature of the ionosphere which until now has been the only alternative. The use of satellites in navigation and position finding also grows rapidly while safety at sea is enhanced by the addition of distress facilities where immediate access to rescue services is available.

Back on land, although most traffic has easy access to land-based telecommunication systems, this does not apply for vehicles travelling long distances between centres of population, in the extreme for example, when crossing a desert. Land mobile satellite communications therefore provide a service for long distance travel, such a service previously hardly existed.

Weather satellites are able to examine the conditions of the earth and the clouds from up above by optical means, in fact global coverage by photography is quite feasible. Threatening weather phenomena such as storms and hurricanes can be recognized in advance of their becoming dangerous. In addition to photography, wind speeds, temperature and moisture content of the air are signalled back to earth to aid in weather prediction. More recent satellites provide meteorological observations in colour from a geostationary orbit.

Military use extends over a wide range. Investment and research in the development of these systems is considerable and it may well be that such use tends to lead the field in advanced, highly developed methods. Of special interest are the small tactical terminals used by the army on land. Such a terminal may be carried by a small vehicle and erected quickly. An even smaller terminal is the "manpack" which can be carried by one man for field use (see

Chapter 8). Optical lenses are now so well developed that a satellite can pass over the earth, photographing the terrain as it goes, then transmitting the pictures back to earth by radio.

1.3 Organizations

By their very nature, satellites which provide telecommunication services on a global basis must be controlled by international organizations capable of arranging agreements as to the division of operating, use and financial arrangements. Clearly also there must be some control as to where satellites are placed. There are also domestic systems, intended for use by a single country yet also available to surrounding countries although usually at a reduced signal strength.

The main international organizations are INTELSAT which has many satellites in orbit, INMARSAT for maritime, aeronautical and land-based mobiles and INTERSPUTNIK which serves mainly Soviet aligned countries. Less international are EUTELSAT (European) and ARABSAT for the Arab state countries. There are many in the domestic category, for example TELECOM (France), ANIK (Canada), TV-SAT (Germany). In fact many countries have their own systems.

1.4 The Electromagnetic Wave

Fundamental to all satellite (and other radio) transmissions is the electromagnetic wave. Right from the start Nature has provided us with a range of these waves in the form of heat and light from the sun. More recently we have discovered how to generate our own waves, not only for heat and light but also for communication purposes. On this modern society relies, in fact take away the radio and other communication systems and we are back over 100 years.

It was Heinrich Hertz, the German physicist who set the ball rolling in the late eighteen-hundreds. He established the basic properties of radio waves which in fact were originally known as Hertzian waves. It was only through his fundamental research that radio communication then got off the ground. In recognition of his pioneering work we now quote wave frequencies in hertz, kilohertz, etc.

The electromagnetic wave is a most complex affair so we cannot expect to understand it fully, in fact nobody does. Some of our difficulties must surely arise from the fact that it cannot be seen and that it moves unceasingly fast. In fact as we read this, the air around us is full of electromagnetic waves speeding past yet undetected by any of the human senses. For proof switch on a radio receiver and we are then picking out just one of them. Some of the main features of the wave are described in the following sections.

1.4.1 Fields

Perhaps the best way of introducing the subject of fields is by discussing first the field of gravity in which we all live. It is all around and even inside us, yet what is it really? Just one more facility Nature has provided but how it works she keeps to herself. That although unseen it has power is obvious for it can pull everything down to earth. Technically the field of gravity affects anything which has mass, which happens to be everything on earth.

There are other fields, for example when a magnet picks up a steel pin, something must have crossed the space between them to seize the pin and draw it to the magnet. We call this something a magnetic field. In the same way an electric field exists in the space between the opposite poles of an electrical supply, i.e. between opposite charges. These fields are difficult to appreciate because like gravity they are invisible, but unlike gravity they only affect certain materials. Perhaps a field is best described as a sphere of influence, i.e. something which has the potential to create some sort of a force. We need to appreciate these facts because the electromagnetic wave on which the whole of modern society rests is basically an electric field linked with a magnetic one, nothing else, simply two interlocked fields travelling together.

We manage to portray a field by the use of a grid of lines arrowed to indicate the field "direction". This again is purely our own way of seeing things. Thus a magnetic field acting in a direction left to right is drawn as a series of arrowed lines (the flux) as shown in Figure 1.2(i). The direction of the field is that in which a free North pole would move (a reminder, this is purely imaginary, a North pole cannot exist on its own). The closeness of the lines may also be used as an indication of the strength of the field.

Equally an electric field may be so depicted, the arrows indicating the direction in which a free positive charge would move (free positive and free negative charges *can* exist).

The electromagnetic wave has its two fields at right angles, hence the wave might be shown on paper as in Figure 1.2(ii). This is at a particular instant. Less than a millionth (or even one millionth of one millionth) of a second later a picture of the same wave might have the arrows reversed as the wave goes via zero through half of one cycle. The two fields are coupled together in such a way that each is a product of the other, a phenomenon with which here we dare not get involved. With a *plane* wave the two fields and the direction of propagation are mutually perpendicular and if the electric field lines are vertical as in Figure 1.2(ii) then the wave is said to be vertically polarized (V). Rotate the figure through 90° then the wave is horizontally polarized (H). There is also circular

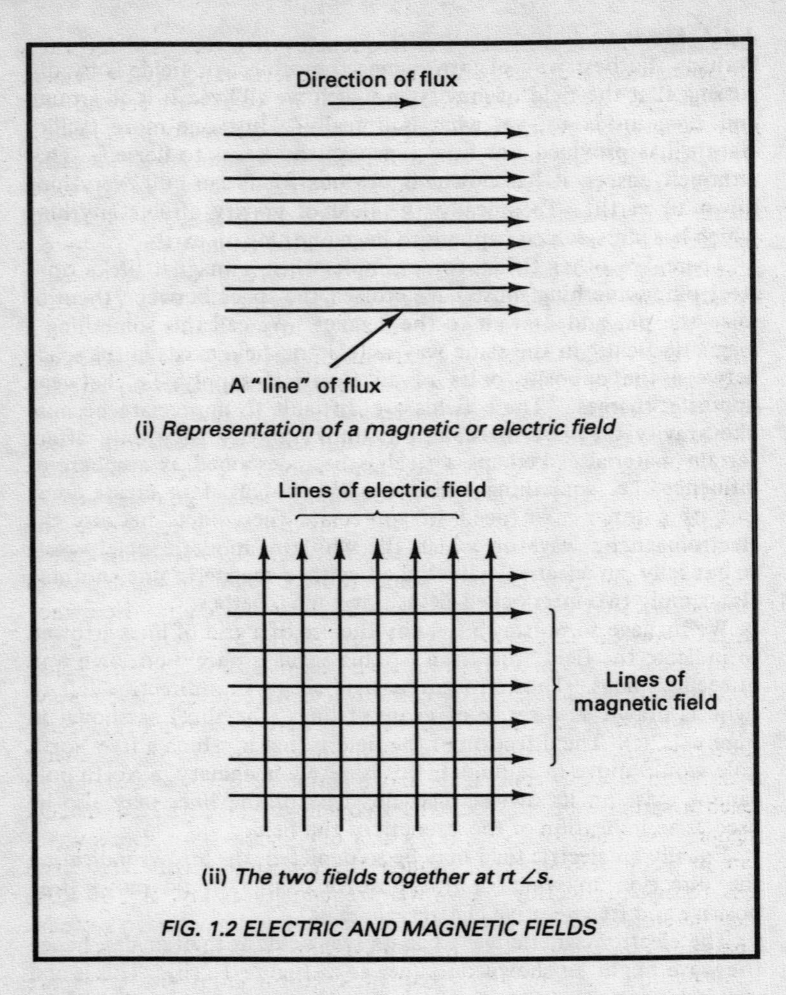

(i) *Representation of a magnetic or electric field*

(ii) *The two fields together at rt ∠s.*

FIG. 1.2 ELECTRIC AND MAGNETIC FIELDS

polarization in which Figure 1.2(ii) can be considered to be rotating. Elliptical polarization is even more difficult to visualize. We consider polarization in greater detail in Chapter 4.

1.4.2 Frequency

The electromagnetic wave changes from one direction of flux, through zero to the opposite direction and back again in one cycle. If this happens once per second, we say the frequency of the wave is 1 hertz (1 Hz). Frequency is therefore the number of complete cycles of a periodically varying quantity occurring in a specified time, usually one second. The frequencies of waves used for satellite transmission are generally in the gigahertz range (GHz) where 1 GHz = 10^9 Hz, i.e. one thousand million cycles each second. We may write these numbers on paper but who can really appreciate a wave changing so rapidly, and it is going on around us all the time!

Knowing the frequency of an electromagnetic wave, the wavelength (λ), that is the distance travelled between two successive similar points of the cycle, can be calculated from $\lambda = v/f$ where v is in metres per second and f is in hertz. For radio waves travelling through the atmosphere or space we use c instead of v (see next Section), hence $\lambda = c/f$ where $c = 3 \times 10^8$ m/s.

An important feature of electromagnetic waves is that they are individuals, two or more waves can exist together with no interference provided that their frequencies are not too close.

We have noted earlier in this Section that satellite transmissions are on much higher frequencies than might be expected considering that terrestrial television for example manages on frequencies of only a fraction of one gigahertz. This is a condition imposed on us when we wish to direct an electromagnetic wave into space. High up around the earth at some 100 to 1000 km are layers of ionized air forming what is known as the ionosphere. Here radiations from the sun provide sufficient energy for electrons of the atmosphere to break away from their parent atoms, the latter then become positive ions and have the effect of bending a radio wave back to earth as though the atmosphere were a radio mirror. Such reflection of waves decreases as frequency rises into the gigahertz region and in fact occurs little above about 1 GHz. Accordingly to reach a satellite up in space the frequencies employed are usually higher than this, for by working at a few gigahertz or above waves can be projected straight through the ionosphere with very little loss through reflection. However frequencies as low as 100 MHz may be employed but generally for weather and navigation systems. The same conditions apply when a satellite transmits back to earth.

Appendix 3 shows how frequencies throughout the range are classified. Generally satellite transmissions are in the S.H.F. range (3 – 30 GHz). Finally one thing for which we must be eternally grateful is that whatever other distortions happen to a wave on its journeys, the frequency does not change. Imagine our tuning difficulties if it did!

1.4.3 Velocity

In any medium the speed of propagation of a wave (v) is controlled by both the permeability (μ) and the permittivity (ϵ) of the medium:

$$v = \sqrt{\frac{1}{\mu\epsilon}} .$$

In free space

$$\mu = \mu_0 = 4\pi \times 10^{-7} \text{ henrys per metre}$$

and

$$\epsilon = \epsilon_0 = 8.854 \times 10^{-12} \text{ farads per metre.}$$

μ_0 and ϵ_0 are constants which we are obliged to introduce so that theoretical calculations can be linked with experimentally observed values. Hence for free space:

$$v = \sqrt{\frac{1}{(4\pi \times 10^{-7})(8.854 \times 10^{-12})}} = 2.998 \times 10^8 \text{ m/s},$$

and for this condition only we substitute c for v. Hence c represents the velocity of electromagnetic waves in free space.

This figure is so nearly 3×10^8 that this is the value generally used. Incidentally free space is simply a theoretical concept. Even when no air or other gases are present, to be truly free there should be no gravitational or electromagnetic fields around, an unlikely condition. Nevertheless the so-called free space is useful as an absolute standard in theoretical calculations as above.

For any medium other than free space, both permeability and permittivity are greater than 1, hence v is lower. However, generally in the earth's atmosphere the velocity is so nearly equal to c that this is the value used.

1.4.4 The Impedance of Free Space

It may come as a surprise to many readers to find that space, that volume of nothingness, actually has an impedance. This follows from the fact that the power (P) residing in an electromagnetic wave eventually gives rise to a voltage (V) and to do this the power must be dissipated in an impedance. From Ohm's Law we recall that $P = V^2/Z_0$. Z_0 is known as the impedance of free space and it has a value of 120π ohms ($377\ \Omega$), it is in fact derived from $\sqrt{\mu_0/\epsilon_0}$ (Sect. 1.4.3). As a single example, if we consider a wave in free space carrying a power of 1 picowatt (10^{-12} watts), then the voltage of the wave is given from $\sqrt{P \times Z_0}$, hence $V = \sqrt{10^{-12} \times 120\pi)} = 1.94 \times 10^{-5} = 19.4$ microvolts. Hence from Z_0 wave powers and voltages can be inter-related.

The impedance of the atmosphere does not differ greatly from that of space.

1.4.5 Coaxial Lines and Waveguides

Although an electromagnetic wave is usually radiated into the atmosphere or space, there are many applications for which the wave must be confined. For this purpose there are two types of transmission line, flexible coaxial cable is employed at frequencies up to just over 1 GHz but above this the loss in such cables may become prohibitive and rigid metal rectangular or circular ducts known as waveguides are used. Such rather complex techniques are necessary because ordinary wires are useless at these high frequencies, they simply disperse the wave energy by radiation. The first requirement therefore is that the wave should be enclosed.

A type of coaxial cable is shown in Figure 1.3(i). The conductors consist of (i) an outer thin copper tube or for greater flexibility, a copper braiding and (ii) a single copper wire running through the centre. The two conductors are therefore coaxial. The centre wire may be held in position by polyethelene discs or by a continuous cellular polythene filling. In both cases the insulation between the conductors is mainly dry air. The radio frequency fields of an electromagnetic wave are contained entirely within the outer conductor hence there is no loss through external radiation nor does the cable suffer from interference from outside radiations.

The characteristic impedance (Z_0) of a coaxial line (i.e. the impedance seen at the end of a long length of the cable or that seen for a shorter length terminated in Z_0) is given approximately by $\sqrt{LC}$ showing that the impedance is more or less independent of frequency. Taking a practical television coaxial cable as an example with inductance, L = 0.3 mH per km and capacitance, C = 0.05 μF per km, then:

$$Z_0 = \sqrt{L/C} = \sqrt{\frac{0.3 \times 10^{-3}}{0.05 \times 10^{-6}}} = 77 \text{ ohms}.$$

This is the result we might expect for generally Z_0 for coaxial cables lies between 50 and 100 ohms.

Waveguides are not so flexible and Figure 1.3(ii) shows a section of a rectangular one. Within the guide the electric and magnetic fields are at right angles to each other and perpendicular to the direction of propagation (along the guide). A close relationship exists between the guide dimensions and the transmission wavelength. As an example a waveguide of internal dimensions 2 × 1 cm might be used to transmit a wave at 12 GHz which has a wavelength of 2.5 cm. Waveguides are sealed and filled with dry air or gas to minimize absorption of the wave by moisture. Because they must

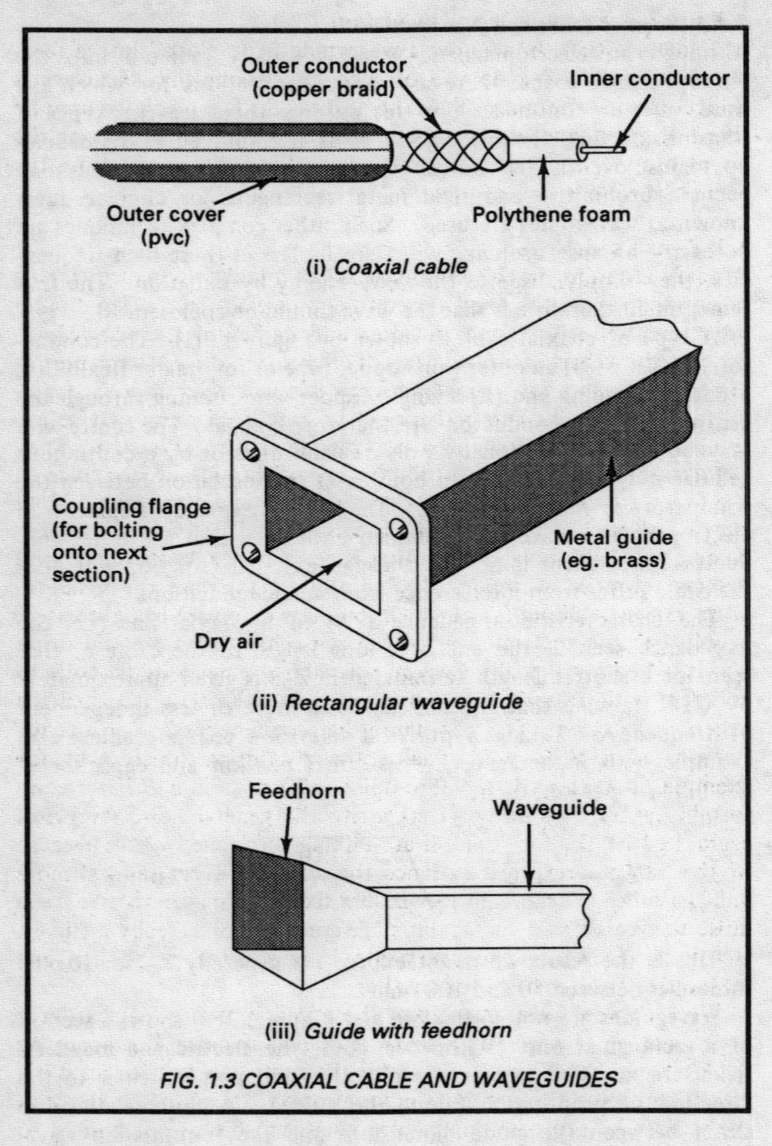

FIG. 1.3 COAXIAL CABLE AND WAVEGUIDES

be accurately machined, are bulky and expensive, waveguides are only used where they are essential. When possible therefore, for flexibility the transmitted frequency is reduced so that the wave can be carried by the cheaper and more flexible coaxial cable. We will see this technique in use in Section 7.2.1.

Figure 1.3(iii) shows how a waveguide is developed into a feedhorn, i.e. with a shaped end of such dimensions that the wave is projected into the atmosphere or space. Such a feedhorn may be fitted at the focus of a parabolic antenna either for projecting the wave towards the dish (transmitting) or for collecting it from the dish (receiving) – see Chapter 4.

1.5 Signals and Noise

A *signal* according to the dictionary is "an intelligible sign conveying information" and the term is used to embrace all electrical and radio intercommunication, from the dots and dashes of the early Morse code to the highly complex affair of colour tv. *Noise* is explained as "irregular fluctuations accompanying but not relevant to a transmitted signal". We can perhaps get to grips with these two definitions by starting with the first two people arriving at a party. They converse and there is no difficulty. A adjusts his or her voice to provide a *signal* which B hears comfortably and vice versa. More guests arrive who chatter among themselves. To either of the first-comers when listening to the other there is now a background *noise* making reception difficult. Technically we say that the *signal-to-noise ratio* (i.e. loudness of signal/loudness of noise) has decreased, simply because the loudness of the noise has increased. In other words, whereas originally the voice signal had practically no noise to compete with, now it has. Of course one alternative is to tell the others to be quiet. This can hardly be recommended so an alternative procedure is adopted which is that by raising the voice, *A*, for example, increases the signal-to-noise (s/n) ratio and *B* hears comfortably again. As more guests arrive the general noise level rises aggravated by the fact that all are talking with raised voices because of the background noise. Hence the s/n ratio decreases still more and the noise is beginning to win. In fact *B*'s ears can receive more noise than *A*'s signal and at this stage conversation is really difficult. Given sufficient noise, conversation is impossible even when one person shouts into the ear of another.

What this demonstrates is that the efficacy of communication is dependent on how greatly a signal exceeds any noise accompanying it. We have shown this to be so for the audio case but in fact it is a general problem throughout communication. Noise is the snake in the grass. To the communications engineer noise is not limited to only that which we hear, any unwanted electrical "irregular fluctuation" is called noise, be it audible or not. As a practical experiment, remove the antenna plug from a television receiver and the screen is immediately filled with white specks. These are due to electrical "noise" generated within the set. A tv picture is spoilt if the incoming signal is not sufficiently strong compared with this internally generated noise.

We can put a few very approximate figures to the noisy party. With a s/n ratio of 16 (signal 16 times louder than the noise), all is well. When the s/n ratio drops to 1, there is difficulty and words get lost (fortunately ears are adept at filling in). However, when the ratio falls to say, $^1/_{16}$ (which is very poor) there is much trouble and little is understood. Working now in decibels for experience we start off with a s/n ratio of 16 which according to Appendix 2 is equivalent to +12 decibels (dB). A ratio of 1 is equivalent to 0 dB and a ratio of $^1/_{16}$ to −12 dB. When the signal power is greater than the noise power therefore, the s/n ratio expressed in decibels is positive, when the signal power is less, the ratio is negative.

Note the use of the word "power". We work mainly in terms of this; "loudness" as considered in the noisy party has meaning only when ears are present.

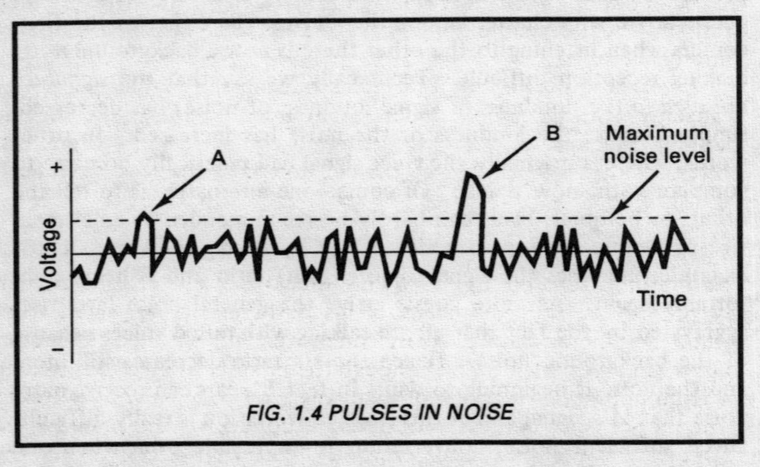

FIG. 1.4 PULSES IN NOISE

How noise can be one of the limiting factors to the usefulness of a communication channel is illustrated by Figure 1.4 which shows how in a channel carrying digital information the pulse at *A* has little chance of being recognized above the general noise level because the s/n ratio is only slightly greater than 1. On the other hand the one at *B* is better off because the s/n ratio appreciably exceeds 1 and the equipment can be arranged to accept only voltages well above the maximum noise level.

The American mathematician, C. E. Shannon, produced a formula which enables us to focus on the main features affecting the transmission of information over a channel. The formula is:

$$C = W \log_2(1 + s/n) \text{ bits per second}$$

where C is the channel capacity, W is the channel bandwidth and s/n is the channel signal-to-noise ratio.

We need not concern ourselves too much with what this all means but the formula certainly confirms that it is not so much the noise level which matters but the degree by which the signal exceeds it. Here we see mathematics endeavouring to get to grips with the multitude of inconsistencies which surround the concept of information flow. Basically what the formula tells us is that if noise on a channel increases, the signal-to-noise ratio decreases hence so does the channel capacity (i.e. the amount of information which can be sent over the channel in a given time). To restore the channel capacity, the channel bandwidth must be increased – this is usually costly.

Summing up – channel capacity is affected by the signal-to-noise ratio and to a certain extent signal-to-noise ratio and bandwidth are interchangeable.

External noise finds it way into a transmission circuit mainly by capacitive or magnetic coupling. Sources of this noise include ignition systems, electric motors, relay contacts, arc welders, etc. Internal noise is generated by current flow in resistances, diodes and transistors.

1.5.1 Thermal Noise

Nature unfortunately has her own source of noise which she unkindly puts in all our equipment. This is thermal noise which arises from the random nature of electron movements within a conductor, i.e. when a current flows. Although the number of electrons passing a given point *on average* is constant, there are momentary variations. The noise voltage measured across a conductor depends on the temperature, hence the description "thermal". The higher the temperature, the greater is the thermal noise voltage, simply because at higher temperatures free electron velocities increase because the added heat provides them with more energy. The effect is known as *thermal agitation* and because these changes occur at random, thermal noise frequencies extend over the whole spectrum.

In many systems, including satellite communication all kinds of noise can be measured and quoted together as an *equivalent noise temperature.* It is evident that this concept is helpful in assessing and comparing noise from different sources. As an example, for the output of an antenna it is the temperature at which the antenna should be if it were to produce the same amount of noise merely as a product of thermal agitation. Equivalent noise temperature is considered in more detail in Appendix 5.

1.6 Digital Signals

Since time immemorial the waveform of speech has been *analogue* (from the Greek, proportional to). It still is, and now we use the term to describe any electrical signal which has a similarity with the original quantity which it represents. The properties such as amplitude, frequency or phase of an analogue signal are therefore variable with time. Audio signals generated by a microphone provide a good example for they vary in all three of these parameters in sympathy with the sound wave which originally generated them.

More recently along came the digital signal (from Latin, *digitus*, a finger), one which can only take certain discrete values, usually few in number but now almost universally, two only. This has given rise to the *binary* (of two) system. Computers and transmission systems are based on the binary digit or *bit*. In electronics generally the two binary states are usually denoted by the numbers 0 and 1, taken from the decimal system. Generally 0 is indicated by no (or very little) voltage, 1 by some voltage of sufficient value to be easily distinguishable from the 0. For transmission systems the near certainty of being able to decide between two states only makes binary ideal for digital work. The binary system is not so new however, it had its beginnings in telegraphy in the eighteen hundreds. Our modern 1 and 0 have their earlier parallels in the *dash* and *dot* of the Morse Code and also in the *mark* and *space* of the telegraph code (in earlier systems a pen made marks or left spaces on a moving paper tape).

A 7-digit code is sufficient to cater for $2^7 = 128$ different characters and commands, e.g. 52 upper and lower case letters, 10 numerals, punctuation marks and symbols. As an example, in one system, the letter D is coded 1000100 while the lower case d is 1100100 with the comma, 0101100. Graphical pictures of digital signals are given later in Figure 5.3.

Channels which are analogue at both ends, e.g. a telephony circuit, are first converted to digital by means of an *analogue-to-digital converter.* This is a device which samples (measures) the analogue waveform at regular intervals and produces a series of pulses corresponding to the voltage level at each time of sampling. Conversion of the pulses back to analogue is via a *digital-to-analogue converter.* Over a complete channel digital transmission has the significant advantage of being able to use *regenerative repeaters* which accept a relatively poor digital signal and transmit onwards a signal identical to the original. This contrasts with analogue transmission where line amplifiers have no choice but also to amplify noise and distortions picked up by the signal on its way. After several amplification stages therefore an analogue signal may be quite different from the original but the digital is "as new". A digital transmission system therefore provides a transmission quality independent of distance.

Chapter 2
GETTING THEM UP

Around the earth there is a layer of gases known as the atmosphere, decreasing in density with height until at about some 100 – 200 km high we might call it space. Any object travelling through the atmosphere has therefore to contend with atmospheric drag which clearly becomes less with height. There is also the inevitable pull of the earth's gravity. In addition the satellite must be released at such a speed that it remains in orbit round the earth at the almost incredible velocity of many thousands of kilometres per hour as developed in Chapter 3. Only a rocket can do all this.

2.1 Rockets

Basically a rocket is propelled by the reaction due to a continuous jet of rapidly expanding gases generated by the combustion of fuel at a high temperature. Newton first put us wise to this for he said that for every action there is an equal and opposite reaction. Take the example of a machine gun, for every bullet fired there is a recoil, i.e. the forward momentum (mass × velocity) of the bullet is equal to the backward momentum of the gun. When firing rapidly the recoil becomes more of a constant push. In the case of the rocket it ejects not bullets but a stream of hot gases and it is the recoil which pushes the body of the rocket upwards. Put more technically, the forward gain in momentum of the rocket must be equal to the backward gain in momentum of the ejected gases. Note that the rocket gases are not pushing against air, it is purely a reaction hence rockets are just as easily propelled in space.

Rockets are massive as we have all seen on our television screens when one lifts off. Tonnes of fuel are burnt every second while the huge contraption slowly rises from the ground. There are two main types of rocket carrier systems:

(1) expendable launch vehicles (E.L.V.). These are unmanned and the various stages of the rocket are discarded when their work is done until finally only the satellite remains. A typical design is shown in Figure 2.1(i), this shows the outline of an Ariane IV rocket. It can only be used once.

(2) the space transportation system (S.T.S.). This is essentially a special aircraft carried up by a rocket. The aircraft carries astronauts who on completion of their mission (e.g. placing satellites in orbit) glide back to earth and finally land as would a normal aircraft. The present one is sketched in Figure 2.1(ii), it is the SPACE

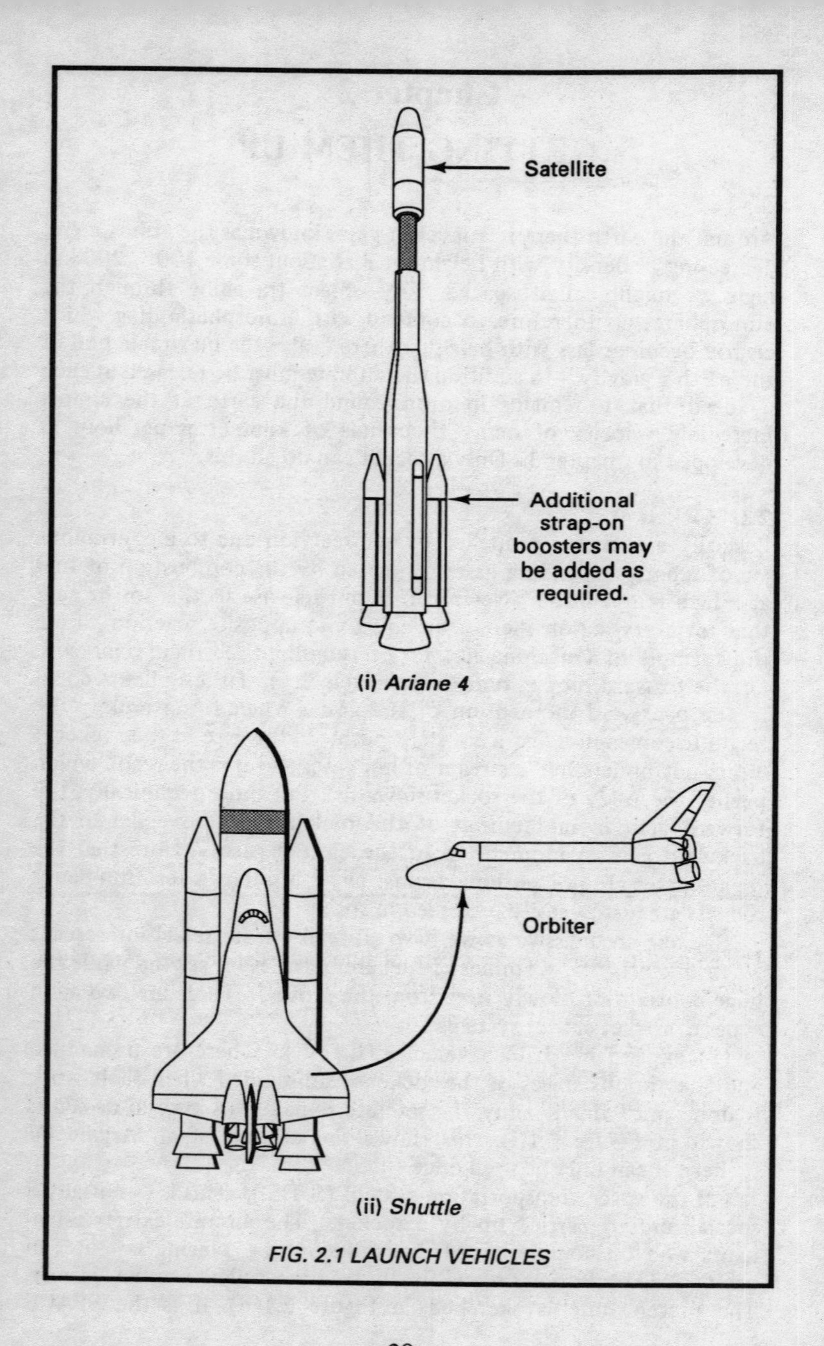

FIG. 2.1 LAUNCH VEHICLES

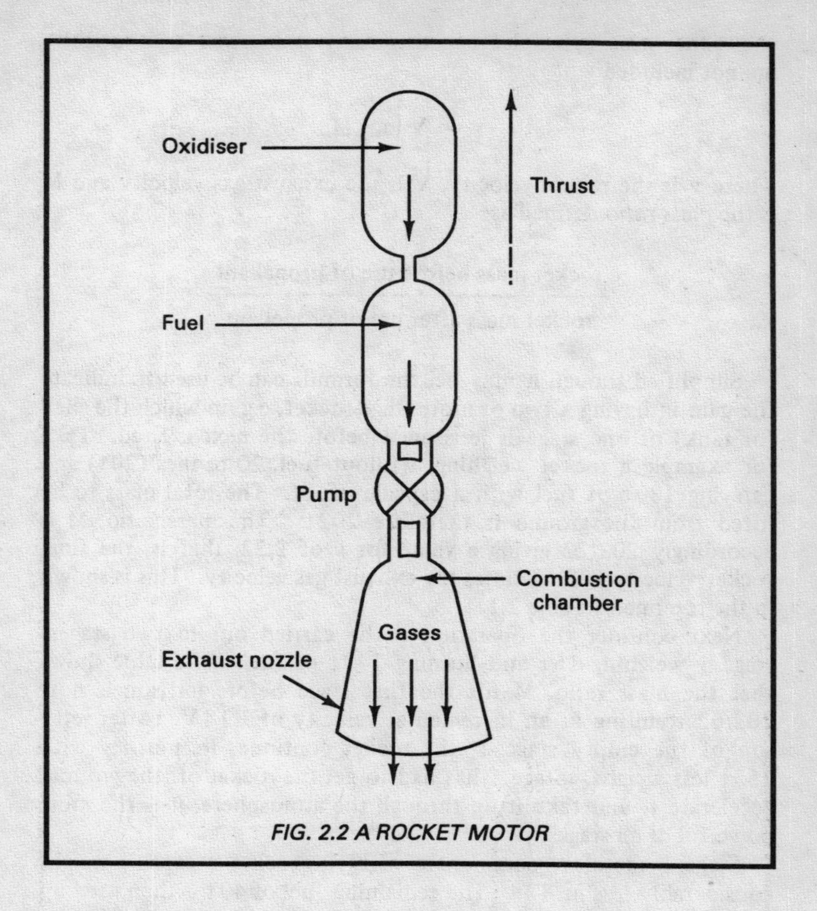

FIG. 2.2 A ROCKET MOTOR

SHUTTLE, as large as a modern jet liner and capable of carrying a payload of nearly 30 tonnes. The rocket is discarded in stages and the shuttle continues its mission steered by rocket motors. The shuttle can be re-used.

The rocket itself contains many tonnes of fuel which is pumped into the combustion chamber as shown in Figure 2.2. Many different fuels are available, some are of highly complex chemical composition but more generally understood are liquid hydrogen for the fuel and liquid oxygen for the oxidiser. The oxidiser (not necessarily oxygen) supplies oxygen for the burning process (oxygen from the air is not available high up).

There is a simple formula which indicates the rocket velocity in terms of the amount of propellant consumed. It can no more than show the basic principles because many unknowns and variables are not included:

$$\nu = V \log_e M$$

where ν is the rocket velocity, V is the exhaust gas velocity and M is the *mass ratio* defined as:

$$\frac{\text{rocket mass before use of propellant}}{\text{rocket mass after use of propellant}}$$

Simplified though it may be, the formula can be used to indicate the gain in having a two or more stage rocket, e.g. in which the shell (or tank) of one stage is jettisoned before the next is fired. Take for example a rocket weighing, without fuel, 20 tonnes (20 t) and carrying 180 t of fuel with a capsule of 2 t. The total mass to be lifted from the ground is therefore 202 t. The mass ratio, M is accordingly 202/22 giving a value for ν of 2.22, that is, the final rocket velocity is 2.22 times the exhaust gas velocity. This is shown in the top line of Table 2.1.

Next consider the operation to be carried out in two stages: stage 1 weighing 15 t and burning 140 t of fuel. The table shows that the mass ratio, M for the first stage before jettison is now 202/62 resulting in an incremental velocity of 1.18 V. After jettison of the empty stage 1, the rocket continues its journey with 155 t less weight. Stage 1 has had to get the rocket off the ground, accelerate it and take it up through the atmosphere, it is the most powerful of all stages.

Stage 2 therefore commences with the rocket + capsule weight considerably less at 47 t. The remaining fuel of 40 t is then used up leaving a mere 7 t of rocket and payload. The mass ratio is now 47/7 resulting in an incremental velocity of 1.90 V as shown. The total incremental velocity is therefore 3.08 V. Changing from single to two-stage therefore results in an increase in the final velocity of 3.08/2.22 = 1.39, i.e. a 39% improvement. Increasing the number of stages to 3 or 4 brings even greater gains.

2.1.1 Launching and Ground Control

Rockets with their satellites attached move fast, in fact the time taken from launch on earth to placement of the satellite in its first orbit is only a matter of minutes. Thus it is all hands on deck when a launch takes place and the total information sent back to the ground control by the rising rocket is considerable. In the opposite

Table 2.1: SINGLE and 2-STAGE LAUNCHING

	Total fuel used (t)	Fuel remaining (t)	Mass (t)		Total	M	Incremental velocity ($V \log_e M$)
			Rocket	Payload			
SINGLE STAGE	180	0	200	2	202	202/22	2.22V
2-STAGE							
Take-off	0	180	20	2	202		
End of Stage 1	140	40	20	2	62	202/62	1.18V
Jettison Stage 1 / Commence Stage 2	140	40	5	2	47		
End of Stage 2	180	0	5	2	7	47/7	1.90V
				Total Incremental Velocity			3.08V

direction the ground control transmits signals to the satellite and its carrier with split second timing. Failure to get the satellite into its appointed orbit is a very expensive affair – and there have been failures! Generally communication from ground to satellite is said to be via the *uplink*, satellite to ground is the *downlink*.

We look at satellite orbits in more detail in Chapter 3 but it is worth noting here that ground to final orbit is not accomplished in a single operation, it is at least a two-stage manoeuvre as illustrated in Figure 2.3. In (i) is shown how the first stage places the satellite in an elliptical transfer orbit round the earth and in (ii) how the second stage moves the satellite into its final working orbit. For the change of orbit an apogee kick motor (AKM) which is a special rocket attached to the satellite is fired at exactly the right time to accelerate the satellite into its final orbit. This may be delayed until the satellite has completed several traverses of the transfer orbit. The apogee is the point in the orbit farthest from the earth whereas the perigee is the point nearest as shown in Figure 2.3(i).

In a launch by the Shuttle the rocket boosters are fired for some 2 minutes whereupon they are detached and then lowered to earth by parachute for re-use. The main rocket then propels the Shuttle to the orbital altitude (some 8 – 10 minutes after lift-off) and the orbiter [see Fig.2.1(ii)] then goes into a parking orbit some 100 km up. Although the orbiter can carry up to around 30 tonnes, it cannot for example deliver this into a geostationary orbit. Satellites destined for this particular orbit must therefore be launched into the perigee of the proposed transfer orbit.

Satellites intended for the geostationary orbit are preferably launched from a site near the equator because (i) the distance to the orbit which is directly above is least and (ii) at the equator the surface velocity of the earth is greatest. In fact while still on the ground we, the rocket and its payload are already travelling in the right direction at over 1600 km/hour (over 1000 miles/hour). This may be a comparatively low speed for a rocket but it is worth having, especially as it is free. One launching site of especial interest to Europeans which is fairly close to the equator is at Kourou in French Guiana from which the ARIANE's go up.

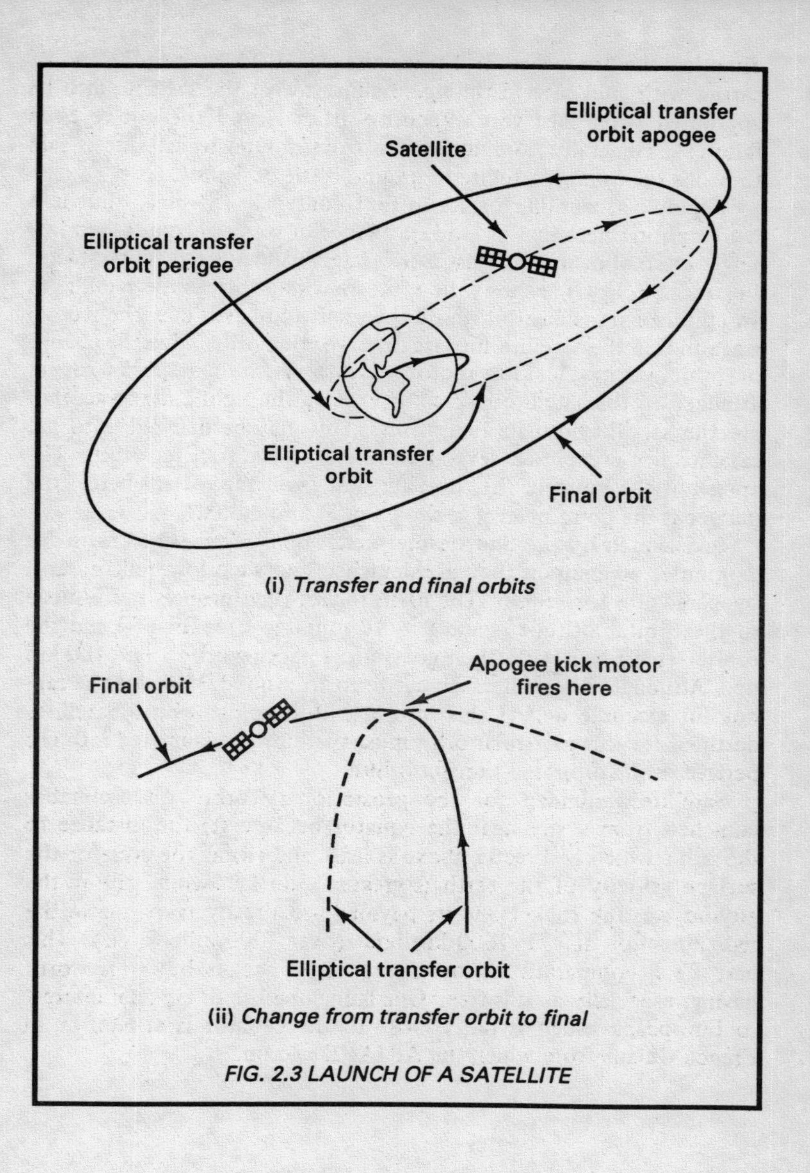

(i) *Transfer and final orbits*

(ii) *Change from transfer orbit to final*

FIG. 2.3 LAUNCH OF A SATELLITE

Chapter 3
STAYING UP

Newton is said to have got his first clue to all this by watching an apple fall to the ground, but this story is not backed up by his writings. It seems rather that he was more concerned with the fact that in spite of the effect of the earth's gravity, the moon did not fall down to earth as other objects did. He soon realized that the moon was moving in its orbit with just sufficient speed to overcome the pull of the earth. In fact the moon is an earth-bound satellite and it stays up because two different forces are in balance. One is gravity, of which we all have plenty of experience, the other is the centrifugal (centre fleeing) force which tends to pull a body away from the centre around which it is moving. Very simply, for a body to remain in orbit round the earth, the force of gravity on it due to the proximity of the huge mass of the earth must be counter-balanced by the centrifugal force due to its mass and speed.

3.1 Circular Earth Orbits

Newton in trying to unravel Nature's mysterious and universal force of gravity found that it attracts an unsupported weight and imposes on it a velocity which increases uniformly, i.e. with an acceleration. It was also discovered both by Galileo and Newton that the force of gravity decreases as the inverse square of the distance. Newton subsequently proposed his Law of Universal Gravity, stating simply that the force of attraction between any two objects is proportional to the product of their masses and inversely proportional to the square of the distance between them. Thus considering the earth as one of the objects with a mass m_e attracting a second object of mass m_s, then

$$\text{force of attraction} \propto \frac{m_e m_s}{r^2}$$

where r is the distance between the centres of gravity of the two objects (i.e. the points at which we can consider the forces to be concentrated).

To relate this to everyday life, a constant (G) is required so that calculations can result in practical results. This constant is now known to be 6.6732×10^{-11} Nm^2/kg^2, not exactly an easily remembered quantity or unit. Hence for a satellite we can write:

$$F_g = \frac{G m_s m_e}{r^2} \qquad (1)$$

where F_g is the earth's gravitational force of attraction on a satellite of mass, m_s. Clearly as the distance from earth increases the gravitational force decreases as one might expect.

For a little more experience we might check first to see whether the correct answer is obtained when the force is at the surface of the earth. The mass of the earth, m_e has been determined as 5.974×10^{24} kg and r becomes the radius of the earth at 6378×10^3 metres. Accordingly:

$$F_g = \frac{(6.6732 \times 10^{-11})(5.974 \times 10^{24})}{(6378 \times 10^3)^2} \cdot m_s = 9.8\ m_s\,.$$

Since force is defined as mass $\times$ acceleration, this is clearly correct because the acceleration due to gravity is known to be 9.8 m/s^2.

Consider next a body moving around the circumference of a circle of radius r. The angular velocity ω is equal to:

$$\frac{\text{angle moved (radians)}}{\text{time taken (seconds)}} \text{ rads/s}$$

from which it follows that $v = \omega r$ where v is the velocity of the body. This is not complete because velocity is a vector quantity (i.e. it has both magnitude *and* direction). In this case the body is continually changing direction from a straight line path to a curved one. Changing velocity implies acceleration and it can be shown that there is a continual acceleration towards the centre of the circle of $\omega^2 r$ or v^2/r. Therefore, from the basic formula:

$$\text{force } (F) = \text{mass } (m) \times \text{acceleration}$$

$$F = \frac{mv^2}{r} \quad \text{newtons} \qquad (m \text{ in kg, } r \text{ in metres})$$

F is the centripetal (centre-seeking) force required to restrain the body from releasing itself from circular motion and flying off at a tangent (i.e. the opposite of the centrifugal force).

Looking at this from the point of view of a satellite:

$$F_c = \frac{m_s v}{r}$$

where F_c represents the centripetal force at a velocity v.

If F_c is taken as the force required to restrain the satellite from releasing itself from orbit and this force is provided exactly by gravity then:

$$\frac{m_s v^2}{r_0} = \frac{G\, m_s m_e}{{r_0}^2}$$

where r_0 is the height of the satellite above the centre of the earth, hence

$$v = \sqrt{\frac{G\, m_e}{r_0}} \qquad (2)$$

and by substituting for G and m_e:

$$v^2 r_0 = 3.9866 \times 10^{14}$$

accordingly if r_0 for any satellite in a circular earth orbit is known, its velocity can be calculated.

3.2 The Geostationary Orbit

In this case we cannot yet calculate v because r_0 is not known. However we do know the time for one revolution in a geostationary orbit which we recall is an orbit vertically above the equator – see Figure 3.1(i). It is not quite the 24 hours as might be expected but slightly less. It is all very complicated because a year cannot be divided exactly into a whole number of days. The earth's journey round the sun is about 365¼ days, we get rid of the odd ¼ by having one extra day each leap year. Unfortunately this is slightly too much so now and again the leap year is omitted (as it was in 1900). However, 365¼ is sufficiently accurate for our calculations. It is known as the solar (of the sun) year and the period of the earth's rotation is called a solar day (24 hours on our clocks).

Not only is the earth rotating on its axis but in a year it also travels once around the sun and as far as the latter is concerned the earth rotates 366¼ times. Accordingly the actual time for a single revolution of the earth is slightly less than 24 hours, in fact it is

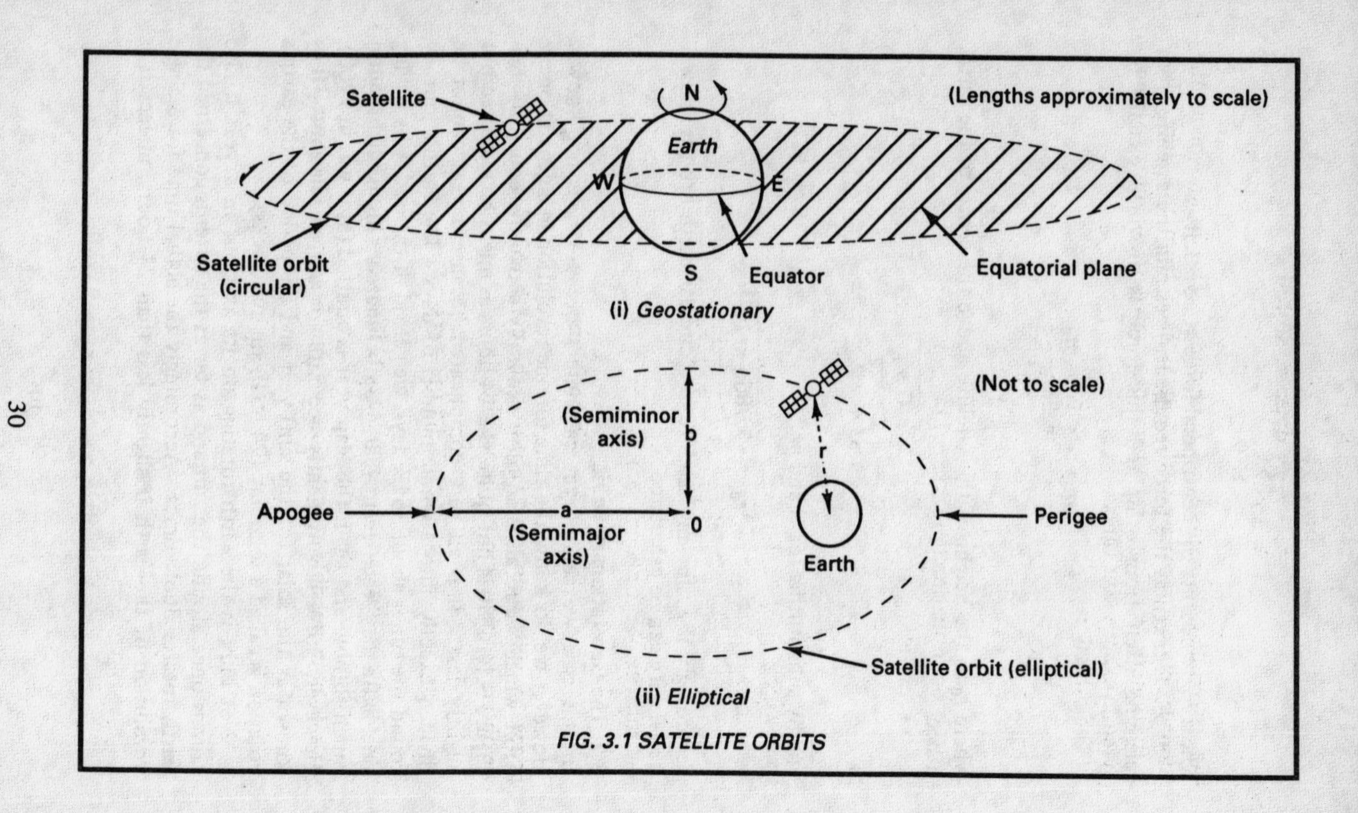

FIG. 3.1 SATELLITE ORBITS

$$T = \frac{365.25}{366.25} \text{ days} = 86{,}164 \text{ s}$$

i.e. 23 hours, 56 minutes, 4.1 seconds. This is known as the sidereal period of rotation, that is, as seen from the stars. For geostationary satellites therefore one revolution must be completed in exactly this time.

The orbital period,

$$T = \frac{2\pi r_0}{v}$$

hence from Equation (2) (Sect.3.1):

$$T = 2\pi r_0 \times \sqrt{\frac{r_0}{\text{G} \cdot m_e}}$$

from which:

$$r_0 = \sqrt[3]{\frac{\text{G} \cdot m_e T^2}{4\pi^2}}$$

and substituting for G, m_e and T:

$$r_0 = 42166.2 \text{ km} ,$$

hence by subtracting the radius of the earth (6378 km) from r_0, we obtain h, the height of the satellite:

$$h = 42166.2 - 6378 = 35{,}788 \text{ km} ,$$

give or take a km or two. In practice G is slightly modified because the earth is not a perfect sphere.

It is now possible to calculate v from Equation (2) (Sect.3.1):

$$v = \sqrt{\frac{\text{G} \cdot m_e}{r_0}}$$

(note here that r_0 is in metres).

$$\therefore v = 3074.8 \text{ m/s} = 11{,}069 \text{ km/h} .$$

Alternatively and perhaps more simply:

$$v = \frac{2\pi r_0}{T} = \frac{2\pi \times 42166.2}{86164.1} = 3.0748 \text{ km/s} = 11{,}069 \text{ km/h} .$$

Summing up, geostationary satellites are situated vertically above the equator 35,788 km high. They circle the earth at a velocity of 11,069 km/h.

3.3 Elliptical Orbits

Clearly the most useful orbit for communication satellites is the geostationary, nevertheless as we have already seen in Section 1.1.2, TELSTAR and the early MOLNIYA satellites used elliptical (12 hour) orbits and such orbits are frequently employed for earth observation and scientific satellites. Elliptical orbits may also be used during the process of placing a satellite in its final orbit – it is certainly not a "straight up and turn left" affair. The elliptical orbit itself needs complicated mathematics for its solution, nevertheless we ourselves can perhaps consider some idealized conditions.

Such an earth-bound orbit is shown in Figure 3.1(ii). Any ellipse has two important points on which it rests, each being known as a *focus.* The orbits are governed by the laws set out as far back as in the 17th century by Johannes Kepler, a German astronomer. He was probably the first to discover that the planets travel round the sun in elliptical orbits with the sun at one focus. One law he gave us states that if the sun is at one focus, a line joining a planet to the sun (which we would call a radius vector) sweeps out equal areas in equal times. From Figure 3.1(ii) therefore it is evident that for this to happen an orbiting planet or satellite must have a changing velocity. This follows from the fact that when a satellite is near the earth it must be moving faster than elsewhere in order to sweep out the same area. This is summed up by the following equation for the satellite velocity in an elliptical orbit with the earth at one focus:

$$v = \sqrt{\mathrm{G} \,.\, m_\mathrm{e} \left(\frac{2}{r} - \frac{1}{\mathrm{a}} \right)}$$

where r and a are the dimensions as shown in Figure 3.1(ii). a and b in the drawing are known as the semimajor and semiminor axes respectively. In this equation the variable is r and as r increases, v decreases.

Another important relationship for an elliptical orbit is that of the eccentricity (ϵ), i.e. the degree to which it deviates from circular:

$$\epsilon = \sqrt{1 - \frac{b^2}{a^2}}$$

with values between 0 and 1.

Hence when b = a as for a circle, the eccentricity is 0, equally when b is very small compared with a, the eccentricity approaches 1.

The time for one revolution (the orbital period) is given by:

$$T = \frac{2\pi a^{3/2}}{\sqrt{G \, . \, m_e}} \text{ hours}$$

where a is in kilometres. A complicated expression indeed. What is more, exact orbit calculations which must allow for other nearby planets and also the fact that the earth is not a perfect sphere, need more than a little help from complex computer programs.

3.4 Power Supplies

Whereas on earth most of the electrical power we use is the result of conversion of some kind of fuel, up in space where there is no filtering by the atmosphere we are more fortunate for the sun provides heat at well over 1 kilowatt per square metre, it is there for the taking. Not all the time for as Figure 3.2 shows, a satellite passes through the earth's shadow once per revolution. This is all very complicated because the sun, earth and the satellite are changing their relative positions continually. The simplest condition when the three are in line is as shown in the Figure.

There are two eclipses to contend with annually, both centred on the equinoxes (about March 21 and September 22 when day and night are equal). The eclipses start at 1 or 2 minutes per day some 20 days earlier, rising to around 70 minutes at the equinox and falling steadily to the few minutes per day over the next 20 days. Hence although there is a generous supply of energy from the sun, the fact that there are interruptions in it, albeit only for short periods, requires the installation of secondary (re-chargeable) batteries to provide the uninterrupted supply of electrical power that a satellite needs. The power required by say, a communication satellite may be as high as 2 kW, although usually less. Space stations manned by astronauts obviously require considerably more. Firstly a sufficient quantity of energy arriving from the sun must be collected, this is by a solar cell array which then charges secondary

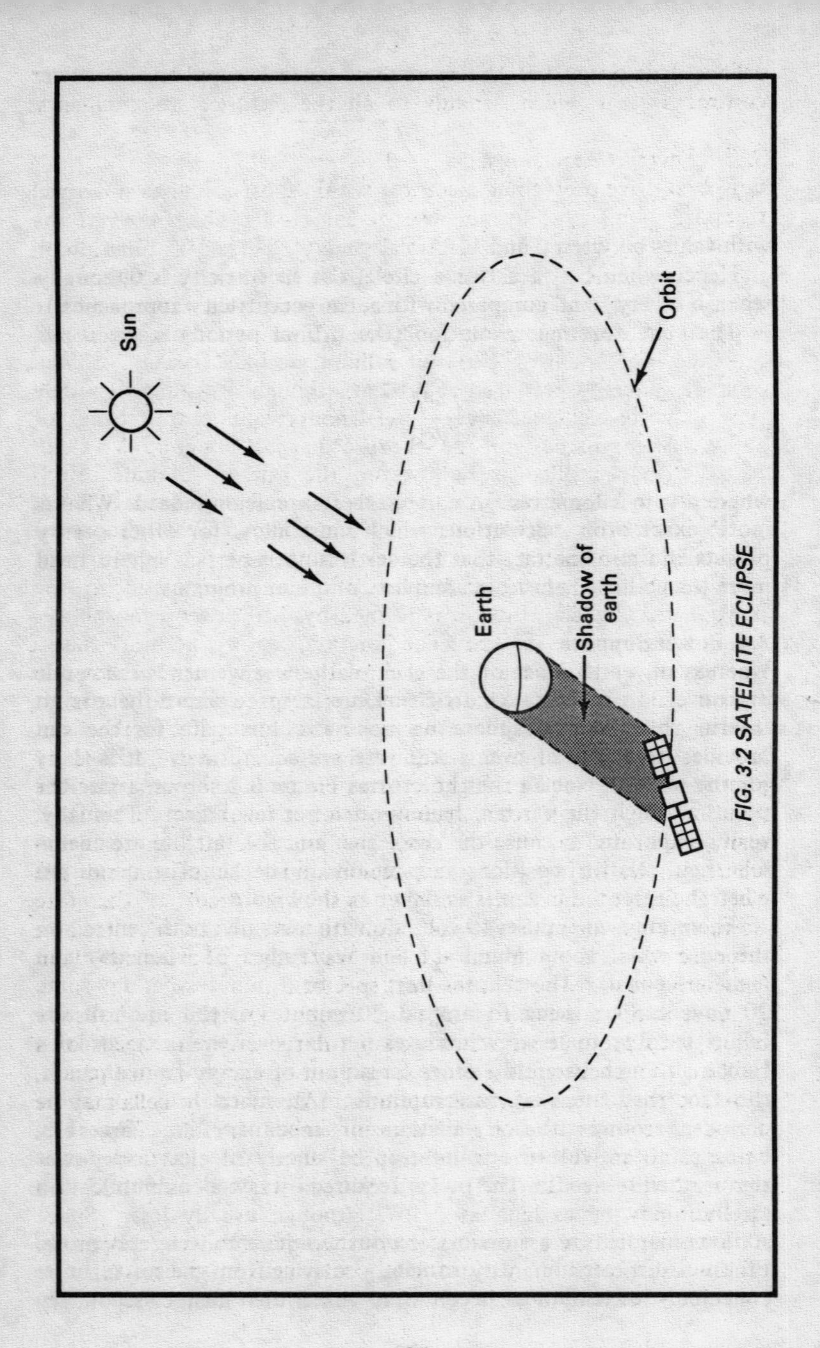

FIG. 3.2 SATELLITE ECLIPSE

cells. The output of the secondary battery supplies the power control system which attends to all the electrical requirements.

3.4.1 Energy Conversion

Power to drive everything electrical within most satellites is derived from the sun's rays by the use of solar cells which convert the primary (sun) energy into electrical energy. These cells function on the photovoltaic effect (electricity from light) which generates a voltage at the junction of two dissimilar materials when these are exposed to electromagnetic radiation. Two types of cell are at present suitable, silicon (Si) and gallium arsenide (GaAs). Silicon cells are generally preferred at present although the gallium arsenide type exhibits higher efficiency and has a greater output e.m.f. of about 1 V compared with the silicon cell at just under 0.6 V. Compared with the silicon cell however, the gallium arsenide cell is heavier and more expensive, it is therefore less frequently used.

The efficiency of conversion of the silicon cell has improved markedly over the last four decades from a mere 4% to at present 15% – 20%. A typical cell consists of a tiny wafer of silicon (the p-type) on which a junction is formed by diffusion of phosphorus (the n-type) on one surface. The junction depth is no more than a fraction of a micrometre. A sketch of the arrangement is shown in Figure 3.3. Connection to the substrate is straightforward but special arrangements need to be made for connection to the front layer so that the minimum area is obscured. Energy released by photons of the incident light creates electron-hole pairs near the junction. Although some recombination takes place continually, many electrons so released cross the junction into the n-region whereas holes diffuse into the p-region under the influence of the electric field of the depletion layer as shown. A current therefore flows through an external load. Construction of the cell aims at a response which is maximum at a light wavelength of around 800 nm (i.e. in the infra-red part of the light spectrum).

At such a low output voltage and current, many silicon cells are required to produce sufficient power for the everyday needs of a satellite. The cells are therefore spread out on flat, wing-like panels, a distinctive feature of most satellites. Alternatively cells may be arranged around the body of a drum-shaped satellite. The cells, being solid, are robust and need to be since they may experience temperatures even lower than –150°C or as high as +60°C in a geostationary orbit.

Generally it is the deterioration of the solar cells which limits the lifetime of a satellite. Unfortunately solar cells in space may be at the mercy of radiation in the Van Allen belts (after James Van

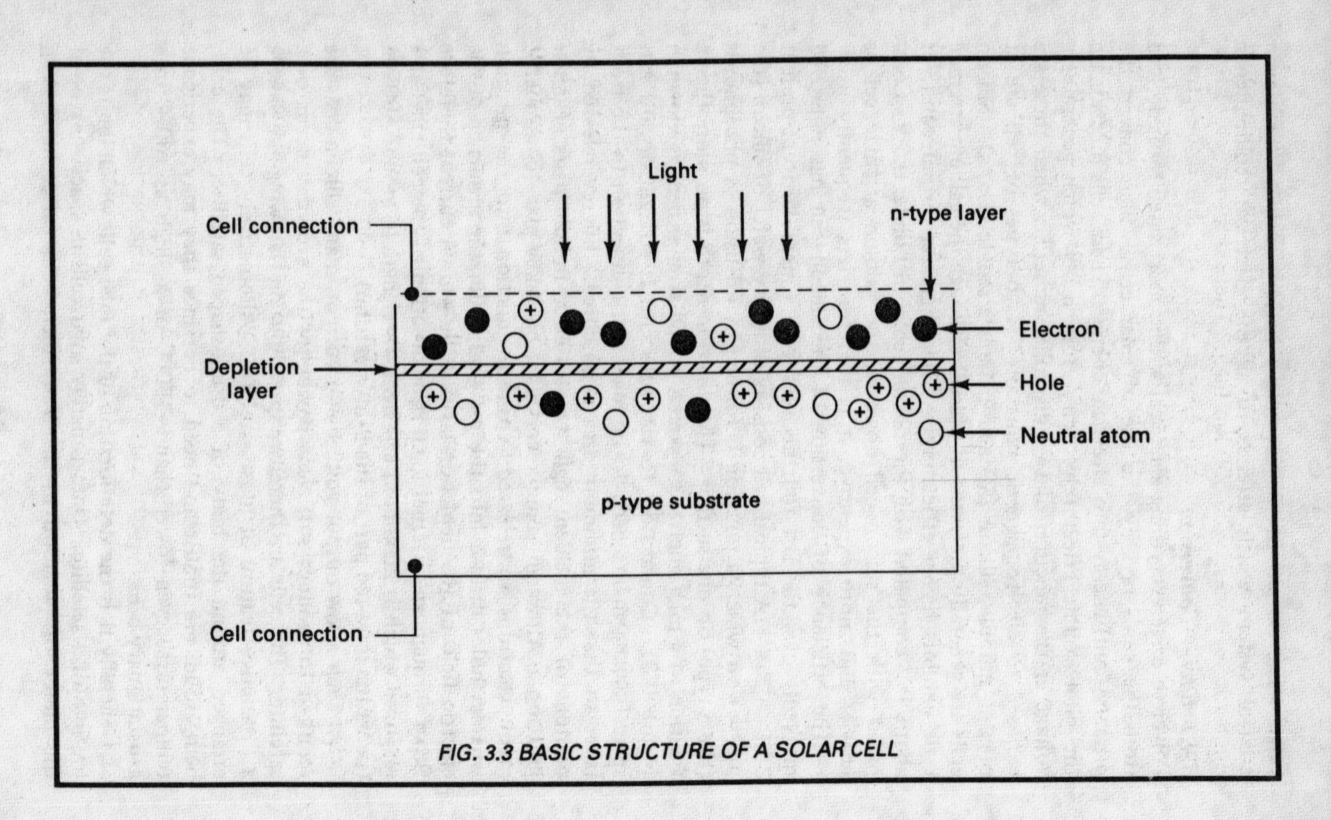

FIG. 3.3 BASIC STRUCTURE OF A SOLAR CELL

Allen, an American physicist). These consist mainly of particles arriving from space forming a belt of free electrons which spiral round the earth nearly 20,000 km high. There is also a belt of free protons which because of their greater mass are able to penetrate deeper into the earth's atmosphere before the earth's magnetic field forces them also to spiral. This however is in a lower belt – a mere 1 – 2000 km high. There is also the effect of solar flares which are accompanied by eruptions from the sun of protons and electrons. All these high-energy particles produce defects in the cell junction, eventually resulting in a reduced power output. Special cover glasses are employed to minimize the effects.

Satellite lifetimes at present are generally around 10 years but as might be expected, all efforts are being made to increase this and now some new satellites have twice this expectation.

3.4.2 Storage Cells

Section 3.4 indicates that secondary batteries must be installed in a satellite to ensure that the electrical power requirements are met during the eclipse periods. The battery capacity is therefore determined from the power required and the duration of the eclipse. Clearly for any battery installed in a satellite the ratio of the electrical energy it stores to its weight should be as high as possible. Lead-acid batteries as used in motor cars are therefore unthinkable.

The cell most favoured at present is the nickel-cadmium (Ni-Cd). This has a positive electrode of nickel oxide with the negative of cadmium. The electrolyte is a solution of potassium hydroxide, resulting in a cell voltage of about 1.25. Capacities are up to 40 watt-hours per kilogram weight (Wh/kg).

A second type of cell is the nickel-hydrogen ($Ni\text{-}H_2$). Compared with the nickel-cadmium, this cell has one of the two metal electrodes replaced by hydrogen gas, this results in a cell having an improved number of cycles of charge/discharge. The cell voltage is 1.3 with capacities up to about 60 Wh/kg. Compared with the nickel-cadmium, a battery of these cells has reduced weight and longer lifetime, however it is slightly bulkier.

The ground control continually monitors the state of the cells and switches them to charge or load as required.

3.5 Position and Attitude Control

Control of the satellite when moving into and also when within its final orbit continually operates via signals transmitted to the satellite by the ground control. The normal antennas used when the satellite is working are highly directional but while the satellite is being dovetailed into its position in the orbit, its orientation with respect to earth may be such that its antennas are not correctly aligned. An omnidirectional antenna is therefore used to receive command signals from the ground control until the satellite attitude has been stabilized. After this the normal directional antennas take over. Antennas are discussed in more detail in Chapter 4.

The satellite attitude (i.e. its orientation relative to earth) must be tightly controlled from the ground. The attitude can be completely described by reference to three mutually perpendicular axes, the roll, pitch and yaw. These are terms also used in shipping and aeronautics and for the satellite, very briefly, the roll axis points in the direction of travel, the pitch axis is normal (perpendicular) to the orbit and the yaw points to the centre of the earth.

The attitude control system must therefore be capable of measuring the satellite attitude relative to what it should be and then signalling this information back to earth. The ground station then determines what corrections are required and transmits the information back to the satellite. These latter signals must therefore rotate the satellite into the attitude required. Such signals are used to control electric motors, rocket motors or gas jets to bring the satellite into the desired attitude, e.g. so that the directional antennas point towards earth and the solar arrays are correctly oriented towards the sun.

The satellite must also be kept in its rightful position in the orbit despite many outside gravitational influences, for although the gravitational pull of the earth is reasonably constant, the attractive forces of the sun, moon and other planets vary.

There are two main attitude control systems:

(1) Spinning – the body of the satellite is cylindrical and spins at up to several hundred revolutions per minute. Normal gyroscopic action opposes any change in the spin axis, hence stabilizes the attitude. With this system the antennas are mounted so that they can be despun so that the direction in which they point remains constant.

(2) Axis stabilization – this employs three momentum wheels which are rotated at high speed to create three-dimensional mutually perpendicular gyroscopic forces (the roll, pitch and yaw axes mentioned above). If the speed of any wheel is changed, the satellite turns accordingly. Hence by controlling the appropriate motor speeds the satellite can be turned into any position.

3.6 Telemetry, Tracking and Command (TT&C)

The title of this section may appear somewhat awesome but in fact it is not. Telemetry means measuring from afar hence the satellite telemetry system makes measurements within the satellite and then transmits the results to ground control. Tracking implies plotting the path of a moving object, e.g. the satellite, while command is the means by which the ground station exercises control.

The telemetry system within the satellite employs many sensors to make the measurements required. From the preceding section it is evident that there must be a group measuring the roll, pitch and yaw errors. As might be expected also, many measurements of the power system are continually made, for example, solar array temperature, voltage and current, battery temperature, voltage and current, current supplied to each satellite sub-system, e.g. gyroscopes, positioning equipment, various pressures and fuel system. Fortunately, although there is a large number of measurements to be made continuously (running into hundreds) most of them vary only slowly with time hence need only a narrow bandwidth signal.

Tracking of a satellite by the ground station requires determination of both direction and distance away. It is usually accomplished by use of a special carrier wave generated in the telemetry system of the satellite. Directional measurements are performed on this wave by the ground station using established direction finding methods. As a single example the range can be measured through phase modulating an uplink carrier by two or more low frequencies. The tones are received by the satellite and on board they are detected and then used to modulate a downlink carrier. The net overall phase-shift in each tone is then measured at the ground station and translated into distance.

Many different command signals are sent up by the ground station to control the satellite, e.g. control of apogee and perigee motors, change of attitude, adjustment of orbit, switching on and off transmitters, receivers and other communications equipment. The decoded command signals are usually digital and within the satellite the received signal is checked for validity against a set of commands stored in a computer memory. It may also be transmitted back to the ground station for further checking. If the ground station finds that the received signal is identical with the originally transmitted command, then an execution command is sent up to the satellite. Only when this procedure is completed does the command take effect.

Chapter 4
ANTENNAS

Fundamental to all radio transmission is the antenna (in the UK formerly known as an aerial). Antennas come in all shapes and sizes but generally the size is related to the wavelength which as it gets shorter therefore requires shorter rods or wires for maximum efficiency. Thus we find that at the lower radio frequencies long wires or metallic rods are employed whereas at the higher frequencies (e.g. UHF – see Appendix 3) the antenna is simply a cluster of short rods, the main element usually being a *dipole*, i.e. with a total length approximating to half the wavelength of the transmission. All very well but at the even higher frequencies imposed on us by the restrictions of the ionosphere (Sect.1.4.2), such rod antennas almost disappear, for example, at 12 GHz a half-wave dipole has a length of a mere 1.25 cm. At SHF frequencies therefore different types of antennas are used, generally parabolic or horn (see Sect.4.2).

4.1 Radiation

Radiation is the emission of energy in the form of electromagnetic waves, in practical terms here it is the process of launching a wave into the atmosphere or space. As pointed out in Section 1.4 it is an extremely complicated affair so we limit ourselves to a few of the basic considerations without going into detail.

Firstly we must accept one of the perhaps lesser known basic laws of Nature which is that an electromagnetic wave is only produced by an *accelerating* charge – it must be accelerating, not just moving. This is not so difficult to arrange in practice for most time-varying electrical quantities involve acceleration. The simplest example is the sine wave for on a graph, from where the curve crosses the axis where the *rate* of electron flow is maximum to its peak where the rate of electron flow is zero, the charge velocity has changed from maximum to zero, hence acceleration is continually involved. It is deceleration in this particular case but mathematically this is considered to be negative acceleration. Radiation therefore increases with the *rate* of variation (the acceleration); at up to a few kilohertz it is negligible, but at gigahertz it is quite a different matter. A transmitting antenna therefore radiates an electromagnetic wave when it is excited by an accelerating current and one of the factors affecting the rate of radiation is the frequency.

The full formula for all this is too awful to contemplate but we can look at a much simplified one from which the electric field arising from an accelerating charge, Q, at a time, t, and distance, d, is:

$$\text{electric field } E = \frac{Q}{4\pi\epsilon_0 c^2 d} \times a(t - (d/c))$$

where a is the acceleration of the charge at the earlier time $(t - (d/c))$ and $c = 3 \times 10^8$ m/s. For ϵ_0 see Section 1.4.3.

Not an easy formula with which to get to grips because both time and distance are affecting E. Nevertheless the formula does show that E is inversely proportional to d. Note that this is when a charge is accelerating, when it is stationary Coulomb's Law shows that E is inversely proportional to d^2. Hence electromagnetic waves can cover large distances and still be useful whereas the effect of a stationary charge is soon lost as distance from the source increases.

Further examination of the formula above will show that t and d are interchangeable, t must increase because time always does, accordingly d also increases, the result being that the electric field moves outwards from the source. A rather sketchy explanation perhaps but we must avoid getting bogged down in complicated theory. As mentioned in Section 1.4.1 the moving electric field carries with it a magnetic field, the two fields being interlocked and inseparable. Either field is capable of inducing a voltage in a receiving antenna and in fact in anything which has conductivity.

4.2 Shapes and Sizes

At the very high frequencies used for satellite transmissions, the antennas seem to take on hitherto comparatively unknown shapes, very different from those used at television and lower frequencies. From the fact that gigahertz frequencies behave in many respects as do light waves we can appreciate that they can be reflected as conveniently as can those of light. The parabolic antenna which is used both for transmitting a wave and for receiving one is therefore to be seen everywhere and contrary to what applies at the lower frequencies, size is not a function of the transmission wavelength but merely of the antenna efficacy, the weaker a received signal is, the larger the diameter of the antenna needs to be. Parabolic antennas are colloquially referred to as "dishes", clearly because of their mildly concave shapes.

4.2.1 The Parabolic Antenna

The foregoing sections in this chapter indicate some of the basic principles of antennas. In this section we look more closely at the salient features of the type of antenna most likely to be encountered in satellite communication, the parabolic antenna or dish. The parabola is a curve well known to all who can recall their school geometry but for those who cannot, the brief notes which follow may help.

A parabola is a particular form of conic section in that the shape is revealed when a cone is sliced in a certain way. It conforms to the equation $y^2 = 4fx$ where f is the point on the principal axis known as the focus as shown in Figure 4.1. Rays such as AB and CD arriving parallel to the principal axis are reflected at the surface of a metal parabolic dish towards the focus. Equally waves launched at

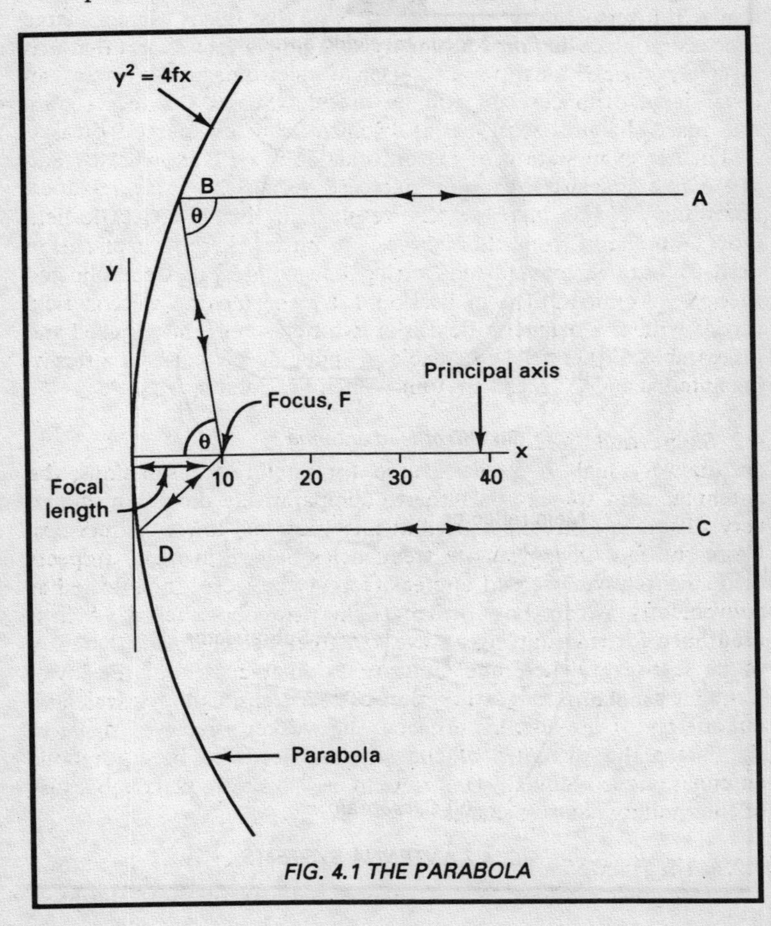

FIG. 4.1 THE PARABOLA

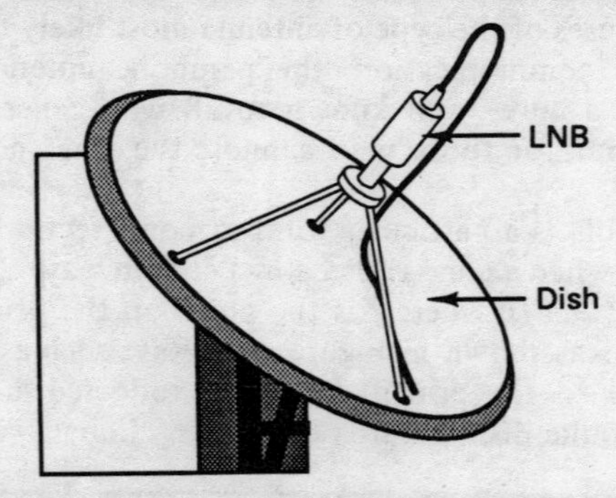

(i) *Prime focus receiving antenna*

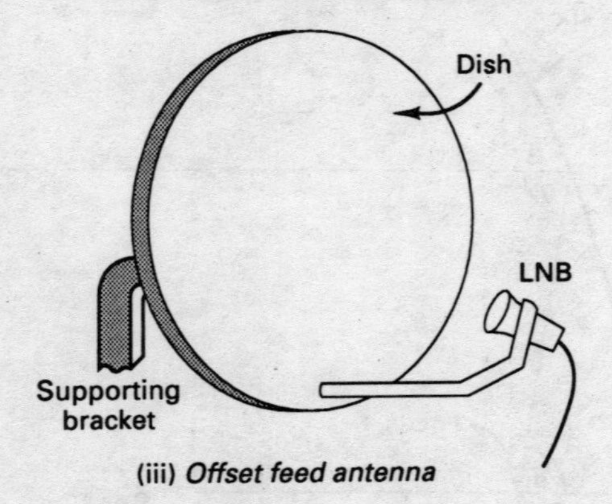

(iii) *Offset feed antenna*

Main reflector

Subreflector

Feedhorn

(v) *Cassegrain*

FIG. 4.2 ANTENNA SYSTEMS

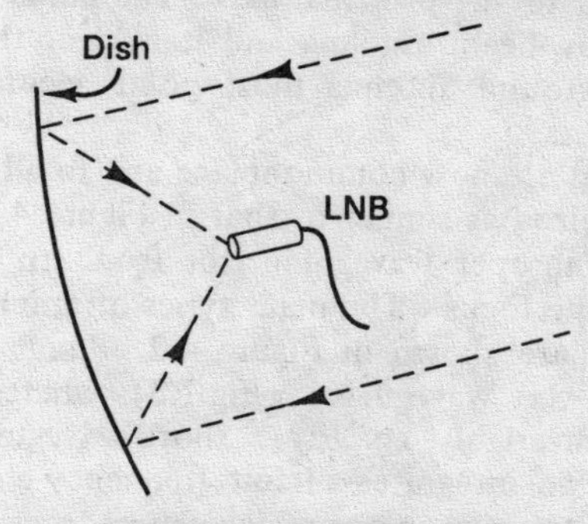

(ii) *Diagrammatic representation of (i)*

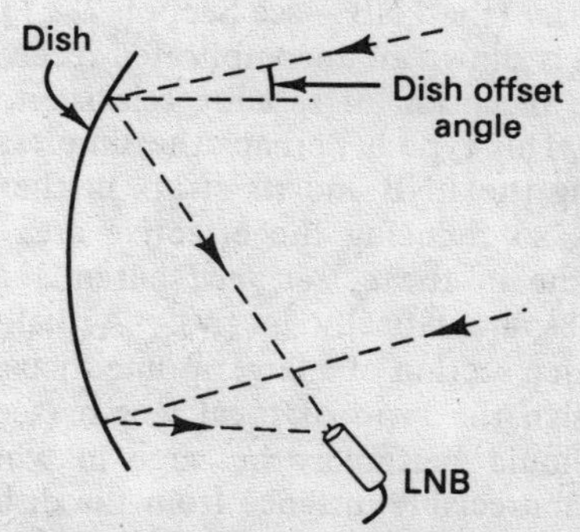

(iv) *Diagrammatic representation of (iii)*

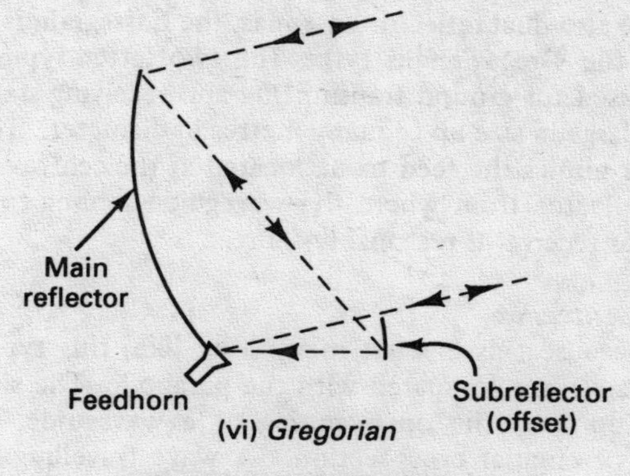

(vi) *Gregorian*

FIG. 4.2 ANTENNA SYSTEMS

the focus and directed towards the parabola are reflected outwards along a path parallel to the principal axis. The parabola therefore functions as a narrow-beam antenna and it is clear that the principal axis of any ground antenna must point accurately to the satellite concerned.

Parabolic antennas are now commonplace and in all of them the shape can be recognized as similar to that in Figure 4.1. They are fed by or feed into an open waveguide [the feedhorn – see Figure 1.3(iii)] located at the focus. The main types of antenna based on the parabolic shape are shown in Figure 4.2. Large transmitting antennas are fed via waveguides (Fig.1.3) whereas domestic receiving antennas invariably employ a short waveguide feedhorn coupled directly to an integrated circuit frequency-changer which reduces the carrier frequency to one which can be transmitted over a coaxial cable (Sect.1.4.5). The device is generally known as a *low-noise block converter* (the LNB – see Section 7.2.1).

In (i) of Figure 4.2 is shown an uncomplicated receiving dish as is used on earth for the reception of satellite television, known as a *prime focus* antenna. This type is perhaps the simplest but has the disadvantage of having the LNB and its struts in the path of the electromagnetic wave, so reducing the effective area of the dish. This effect is overcome in the *offset feed* antenna illustrated in Figure 4.2(iii) and diagrammatically in (iv). Actually the offset antenna is derived from a section of a larger prime focus antenna. In parabolic antenna design the ratio of focal length (see Fig.4.1) to the diameter (f/d) should preferably be large in which case the LNB would be at an appreciable distance from the dish itself. This is generally inconvenient hence a short f/d than the optimum is usually adopted.

There are also dual reflector antennas, the Cassegrain is illustrated in (v) and the Gregorian in (vi). The two latter types are most frequently used for ground transmitting and receiving stations, they are usually large in size up to many metres in diameter. The use of a sub-reflector allows the feed to be located at the centre of or below the main reflector from where the waveguide cabling to the transmitter and/or receiver is reasonably short.

4.2.2 Horn Antennas

So called because this is what many look like, this type provides wide beam coverage compared with the parabolic. The simplest can be looked upon as the open ending of a waveguide, usually of rectangular or circular cross-section. A wave travelling within the waveguide is therefore launched into the surrounding air or space via the horn; in the opposite direction a wave arriving at the horn is

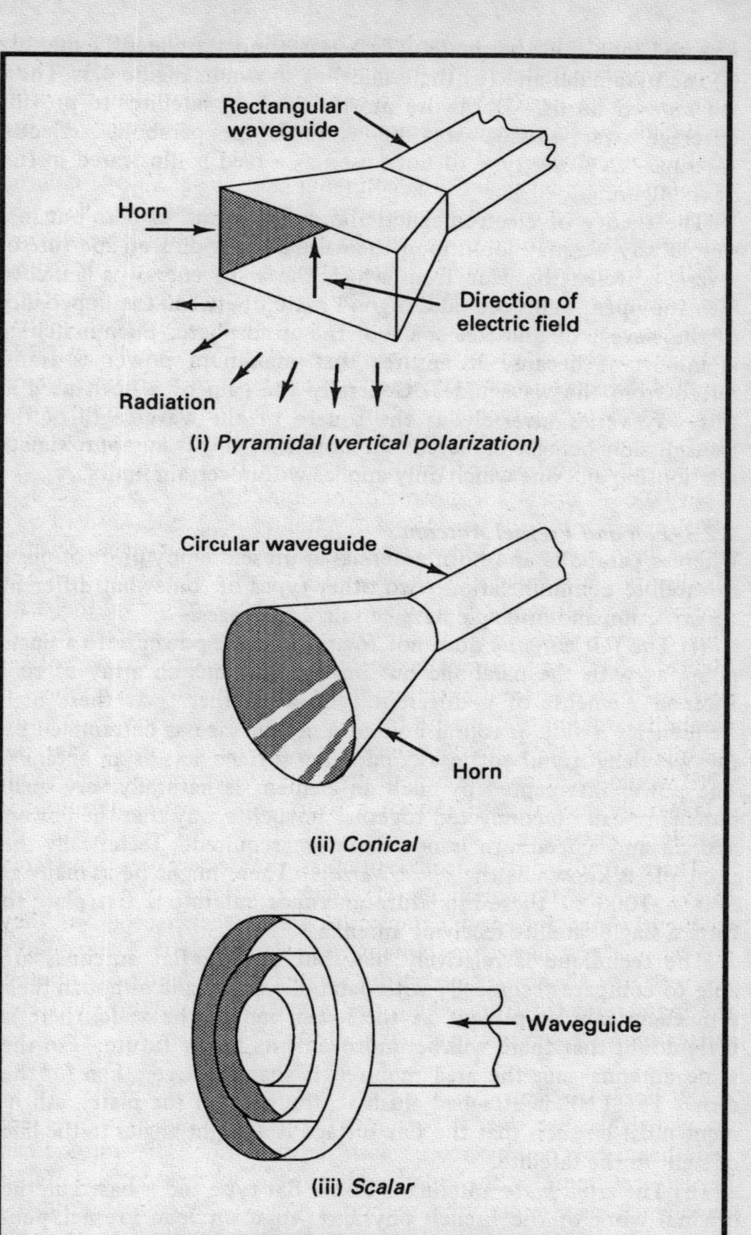

FIG. 4.3 ELECTROMAGNETIC WAVE HORNS

directed into the waveguide. Two types most frequently met are (i) the pyramidal and (ii) the conical as shown in Figure 4.3. These are *tapered* horns. Horns are mostly used on satellites to provide coverage over a wide area or as feeds for parabolic reflector antennas. A *scalar* type of horn used as a feed is illustrated in (iii) of the figure.

The theory of electromagnetic horns is beyond us here but in a simple way we may look upon their use as providing an aperture of several wavelengths wide from which the wave energy is launched into the open. This provides a good match between the impedance of the waveguide and free space or the atmosphere. Such matching is important because it ensures that maximum power is transferred from the waveguide. Generally the gain of a horn used in this way varies inversely as the square of the wavelength of the transmission being conducted but note that this is an approximate relationship and one which only applies within certain limits.

4.2.3 Flat and Fresnel Antennas

Whereas parabolic and horn antennas at present enjoy pride of place in satellite communication, two other types of somewhat different construction and principle have certain advantages.

(i) The *flat antenna* does not focus the wave power onto a single point as with the parabolic but instead contains an array of tiny antenna elements of a different type altogether. At these high frequencies a slot or round hole of a certain size as determined by the wavelength and cut in a conducting surface acts as an antenna. The power intercepted by such an element is naturally very small but many can be connected together in such a way that the powers add up and a feedhorn is not therefore required. Technically the principle is known as the *phased array*. There might be as many as 500 – 1000 of these miniature antennas cut into a flat plate to form a single satellite receiving antenna.

The technique is relatively new but already flat antennas are able to compete technically with parabolic dishes and although their efficiencies are at present at the lower end of the scale, there is little doubt that there will be improvements in the future. For the same antenna gain the area required is slightly larger than for the dish. The LNB is attached flush to the back of the plate. Alignment must be such that the flat surface is at right angles to the line of sight to the satellite.

(ii) The *zone plate* antenna is also a flat type and is based on the original work of the French physicist Augustin Jean Fresnel, published as far back as in the early 1800's. He was one of the first to demonstrate the wave theory of light.

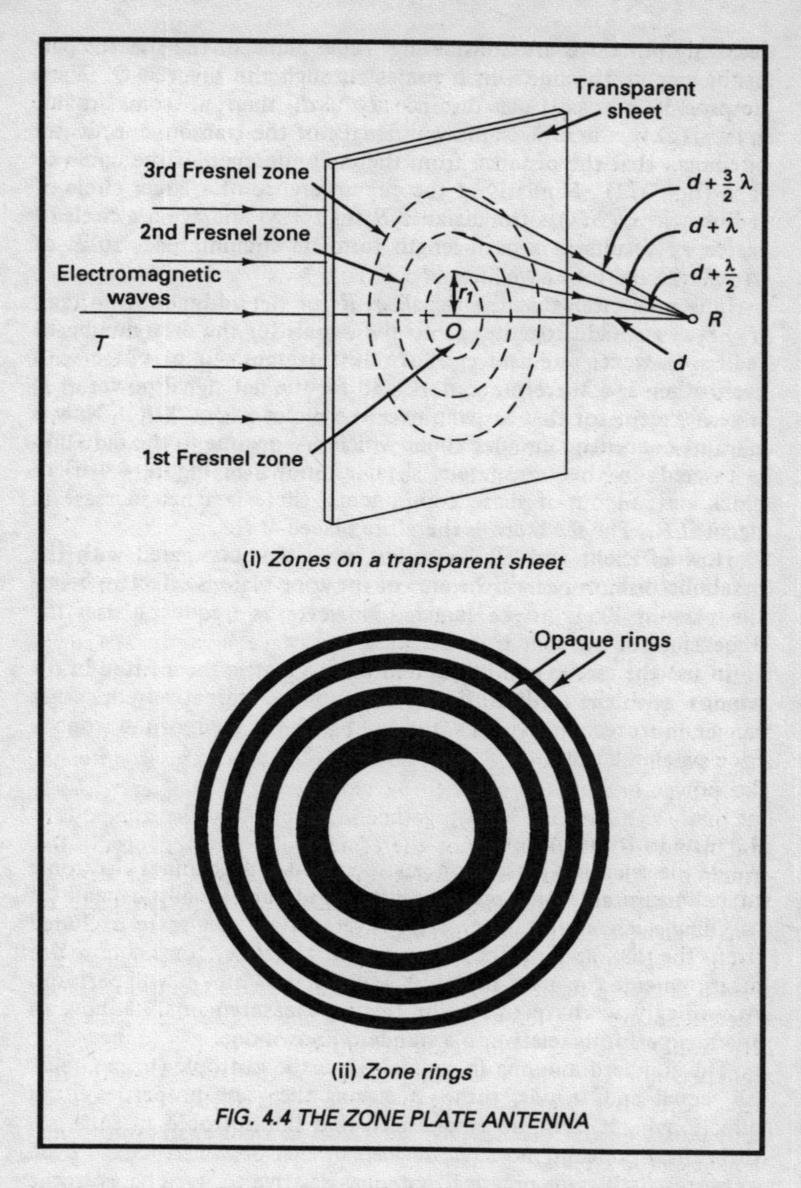

(i) *Zones on a transparent sheet*

(ii) *Zone rings*

FIG. 4.4 THE ZONE PLATE ANTENNA

The principle of the zone plate can be appreciated from Figure 4.4(i). Consider a source of electromagnetic waves at the point T with radiation passing through the transparent sheet. R is the receiving point we are considering. The plane of the sheet is perpendicular to the line which passes through the sheet at O. Very simply, if we call the distance OR, d, then at some radius r_1 [= $\sqrt{(d\lambda)}$] where λ is the wavelength of the transmission, it can be shown that the distance from the circumference of the circle to R is $(d + \lambda/2)$. Similarly at the circumference of a larger circle of radius r_2 [= $\sqrt{(2d\lambda)}$], the distance R is $(d + \lambda)$ and again a circle of radius r_3 results in a path length from its circumference to R of $(d + 3\lambda/2)$ and for r_4 we have $(d + 2\lambda)$.

It is now clear that the signals at R for the odd-numbered radii $(r_1, r_3, \ldots)$ add together as do the signals for the even-numbered radii. However odd and even are 180 degrees out of phase with each other and therefore cancel. So far the net signal power at R is zero except for that arriving over the direct path (TOR). Now if a set of concentric circular zones which are opaque to the radiation is printed on the transparent sheet as shown in Figure 4.4(ii) to block out all out-of-phase components, there is a net increase in signal at R. The feedhorn is therefore placed at R.

How efficient this all looks but note that compared with the parabolic dish, only half the area of the zone plate is effective hence the plate is likely to be larger. However as frequency rises the dimensions of the zone plate decrease.

In use the circles can be printed onto a plastic sheet fitted into a window with the feedhorn inside the room. Alternatively the sheet can be mirrored on the back and used with the feedhorn in front as for a parabolic dish.

4.3 The Isotropic Antenna

We of the scientific persuasion have come to expect things electronic to be measurable and modern instruments are usually capable of fulfilling such expectations. However certain items are excluded from the list and of these the antenna is one, for no *absolute* measurements can be made on it which tell us how it will perform. Accordingly we have to rely on relative measurements, i.e. how an antenna performs relative to a standard known one.

The standard antenna in general use is the isotropic (from Greek, *iso*, equal and *tropos*, turn, i.e. having the same properties in all directions). It must be emphasized that the isotropic antenna is a theoretical concept, no such device can ever be constructed. Even so we can still rate practical antennas relative to it. The isotropic antenna is one which is assumed to radiate from or receive at a single point and be effective equally in all directions as developed in Figure 4.5.

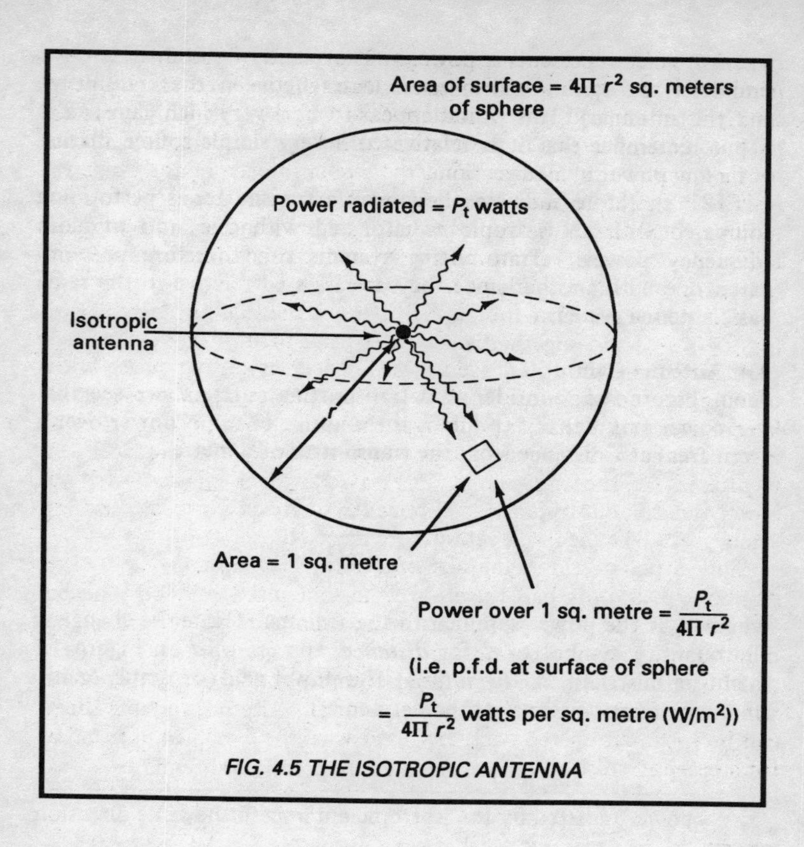

FIG. 4.5 THE ISOTROPIC ANTENNA

4.3.1 Effective Isotropic Radiated Power

This term, frequently referred to simply as EIRP, arises in calculations involving transmitters working at satellite frequencies. It simply expresses the product $G_t P_t$ where G_t is the practical antenna gain in a specified direction relative to the isotropic and P_t is the power supplied to it. If the power radiated is expressed in watts then the EIRP shows how well the transmitter performs relative to an isotropic source. Thus if for example a transmitter output power (P_t) of a satellite is quoted as 40 watts, i.e. approximately +16.0 dBW (note that dBW indicates a power ratio with respect to 1 watt – for decibel notation generally see Appendix 2) and the antenna gain in a certain direction (G_t – see also next Section) is quoted as 30 dB, then:

$$\text{EIRP} = G_t P_t = 30 + 16 = 46 \text{ dBW}$$

which in fact represents a power gain of nearly 40,000. (Here for simplicity we ignore transmission losses between the transmitter and the antenna.) This would appear to be a very high gain and it is but remember that it is relative to a very simple source sharing out a low power in all directions.

EIRP therefore indicates how well a transmitter is performing compared with an isotropic radiator fed with one watt of radio frequency power. Transmitting systems can therefore be compared one with another since they are assessed relative to the same basic antenna system.

4.4 Antenna Gain

From Figure 4.5, considering a transmitting antenna, we see that the power flux density (p.f.d. – the amount of radio power over a given area) at a distance from the transmitter of d metres:

$$\text{p.f.d.} = \frac{P_t}{4\pi d^2}$$

where P_t is the power supplied to the antenna. (Note the change in the quantity symbol to d for *distance*, the use of d on Figure 4.5 might be mistaken for *diameter*.) The power gain (or simply 'gain') of any practical antenna is then defined as:

$$G_t = \frac{\text{power radiated by practical antenna}}{\text{power radiated by the isotropic antenna in the same direction}}$$

most frequently expressed in decibels. Note that here we are considering solid, dish-type antennas, not wire. We will see below that the area of the antenna is all-important.

For a transmitting antenna therefore with a gain G_t, the radiated p.f.d. at a distance away of d metres is:

$$\text{p.f.d.} = \frac{G_t P_t}{4\pi d^2} \text{ watts per sq. metre (W/m}^2\text{)} .$$

This radiated field excites a receiving antenna at the distance d, producing at the antenna terminals a power P_r such that:

$$P_r = \text{p.f.d.} \times A_{eff}$$

where A_{eff} is known as the effective area of the receiving antenna. "Effective area" needs some explanation. It is based on the physical area (A) of the antenna and the antenna efficiency (η), i.e.

$$A_{eff} = \eta A .$$

Antenna efficiencies range from around 0.5 up to nearly 0.8 (see Appendix 7). Values are appreciably less than 1 because (1) the profile of the dish may not be accurate, (2) there is a spill-over of energy at the rim, (3) there may be struts and equipment in the path of the wave.

A universal antenna constant has been derived relating antenna gain G relative to A_{eff} (we will not prove this, things are getting complicated enough). It is:

$$G = \frac{4\pi A_{eff}}{\lambda^2}$$

from which we see that G varies directly with A_{eff} as might be expected. Hence, depending to a certain extent on the antenna efficiency, η, high-gain dish antennas have large diameters.

If the dish is circular, its area is approximately $\pi D^2/4$, where D is the diameter and changing to frequency in preference to wavelength:

$$G = \frac{\pi^2 D^2 f^2 \eta}{c^2} \tag{1}$$

with D in centimetres, f in GHz and η as a percentage this becomes:

$$G = 0.00011 D^2 f^2 \eta \tag{2}$$

Let us take an example from a practical transmitting antenna of diameter 300 cm working at a frequency of 3.95 GHz with efficiency 0.58, then:

$$G = 0.00011 \times 300^2 \times 3.95^2 \times 58 = 8959$$

or in decibels:

$$10 \log 8959 = 39.52 \text{ dB} .$$

Effectively this indicates that the antenna gain relative to the isotropic is 8959 times greater but note that we are considering a fairly large parabolic dish concentrating all its output into a narrow beam whereas an isotropic antenna delivering the same power spreads it in all directions, most of the power therefore being lost.

The above example is in terms of a transmitting antenna, the principles also apply to receiving antennas for which the gain relative to the isotropic is assessed similarly.

In Section 5.2.5 Equation (2) is used for calculation of the gain of practical parabolic dishes.

4.5 Polarization

The concept of wave polarization is introduced in Chapter 1 in which Figure 1.2(ii) shows how a vertically polarized wave might be depicted. Polarization of a wave is generally described by the disposition of the electric field lines, that of the magnetic lines follows automatically since the electric and magnetic fields are at right angles.

As we have seen, the electromagnetic wave defies the imagination especially because it is unseen, travels very fast and generally its power is almost next to nothing. However we may gain some appreciation of what happens from Figure 4.6 which portrays the electric field strength as it changes over one cycle.

Normally the electric and magnetic fields lie in a plane transverse to the direction of propagation. If they maintain their relative directions, e.g. the electric field always vertical, then the wave is said to be *linearly* polarized. One bonus from the fact that we can use either vertical or horizontal polarization is that the same frequency can be used for completely different transmissions provided that they are on opposite polarizations although just to be on the safe side they are usually kept well apart. As mentioned in Chapter 1, which of the two polarizations is effective for a particular transmission is indicated by a *V* for vertical and *H* for horizontal. For any particular satellite channel therefore not only must the frequency be known but also the polarization. The receiving dish LNB, if not fixed to receive one type of transmission only, must be adjusted to match the polarization and rotated (usually by remote control) by 90° for the alternative one. However many types of feedhorn are available with a *polarizer* or *polarotor* built in. Change from one polarity to the other is effected by rotating a flexible membrane within the waveguide. Switching between the two polarizations is by an electromagnet remotely controlled from the indoor receiver.

There is also circular polarization, this is more complicated but can be envisaged as a particular combination of two equal linearly polarized waves continually rotating. Looking towards the

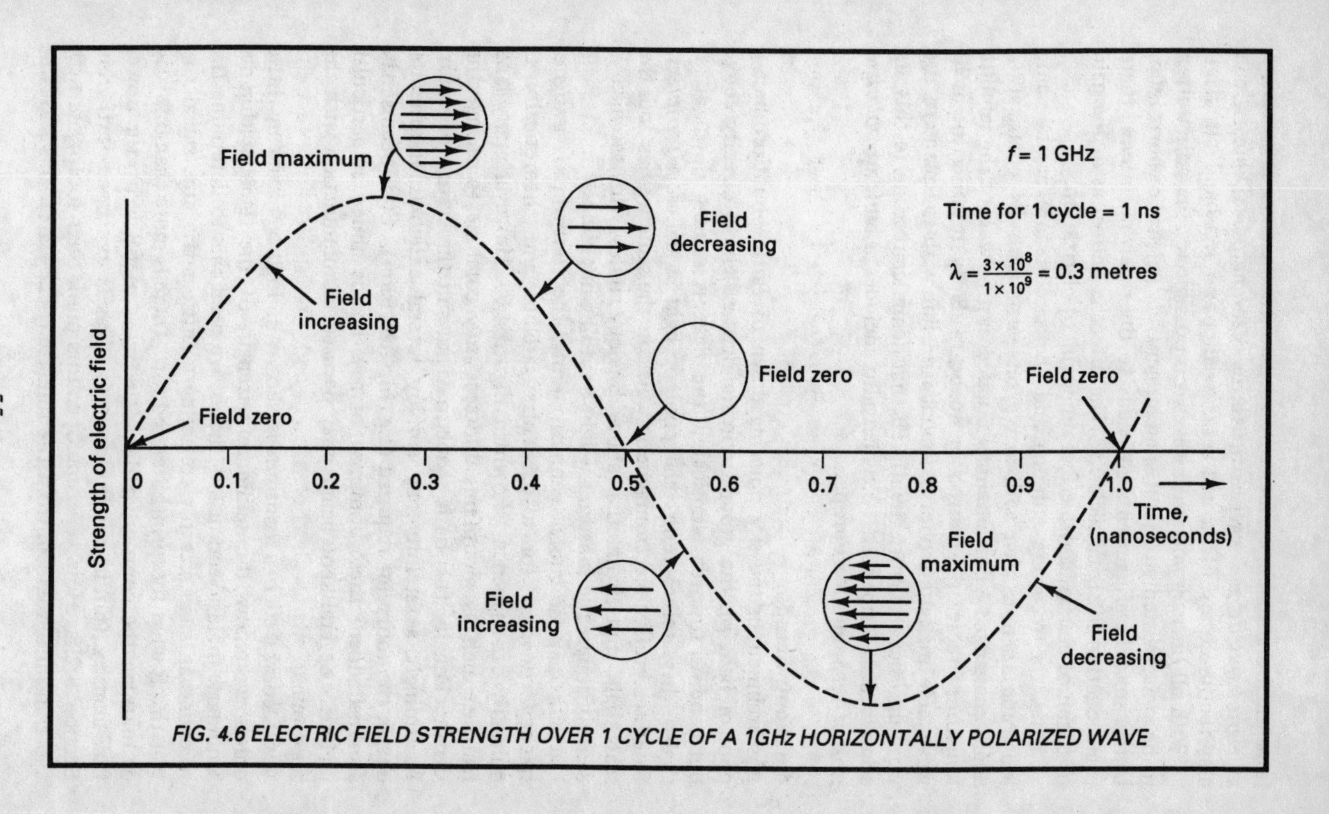

FIG. 4.6 ELECTRIC FIELD STRENGTH OVER 1 CYCLE OF A 1GHz HORIZONTALLY POLARIZED WAVE

transmitting antenna, a *clockwise* rotation of the wave is described as *right-handed*, and *anti-clockwise* as *left-handed.* Labels which may be used are RHC and LHC (right- and left-hand circular).

With all linearly polarized waves it is possible for the polarization to become twisted so that it is no longer truly vertical or horizontal. Unless suitable adjustments are made to the receiving antenna there is a reduction in signal pick-up. The effect is known as *depolarization*, this of course has no effect on circular polarization.

When the receiving antenna and the satellite are at the same longitude then a wave will arrive on earth with its polarization angle unchanged. For antennas situated east or west of the satellite longitude, there is a correction to be made by rotating the LNB waveguide pick-up probe so that it is in line with the plane of the incoming signal electric field. The mathematics relating to this are contained in Appendix 12 which also includes a table of *polarization offset* values for Europe.

4.6 Footprints

A searchlight in the sky pointing down to earth would produce a pool of light on the ground, circular if the light is directly above but tending towards elliptical if the light is shining down at an angle. In satellite terms the pool of light is called a *footprint.* Satellites project electromagnetic waves in the same way as does the searchlight but from the above analogy unless the satellite is directly above, the footprint will be anything but circular.

On a map, in more technical terms, the footprint is a *signal strength contour* for a particular satellite and theoretically is normally egg-shaped as shown in Figure 4.7. However many land features such as mountains and seas may combine to change the shape. Equally this basic shape may not suit the system operator. Accordingly a satellite may employ several antennas in order to obtain the footprint required (e.g. for television). Spot beams may also be added, these comprise narrow beams aimed at particular regions. Several spot beams may be used in combination with the main ones.

Of course such a contour as shown in Figure 4.7 means little unless we know the minimum strength of the signal within its boundary for as Section 4.4 shows, we need this to determine the receiving antenna size for collecting a sufficiently large signal. In fact this is what footprints are used for. There is more than one way of marking the contour, perhaps the most useful is by the power flux density (p.f.d.) on the ground. This is explained further in Section 5.2.5. However some operators mark their footprints with the transmitted e.i.r.p. from the satellite (Sect.4.3.1), this leaves us

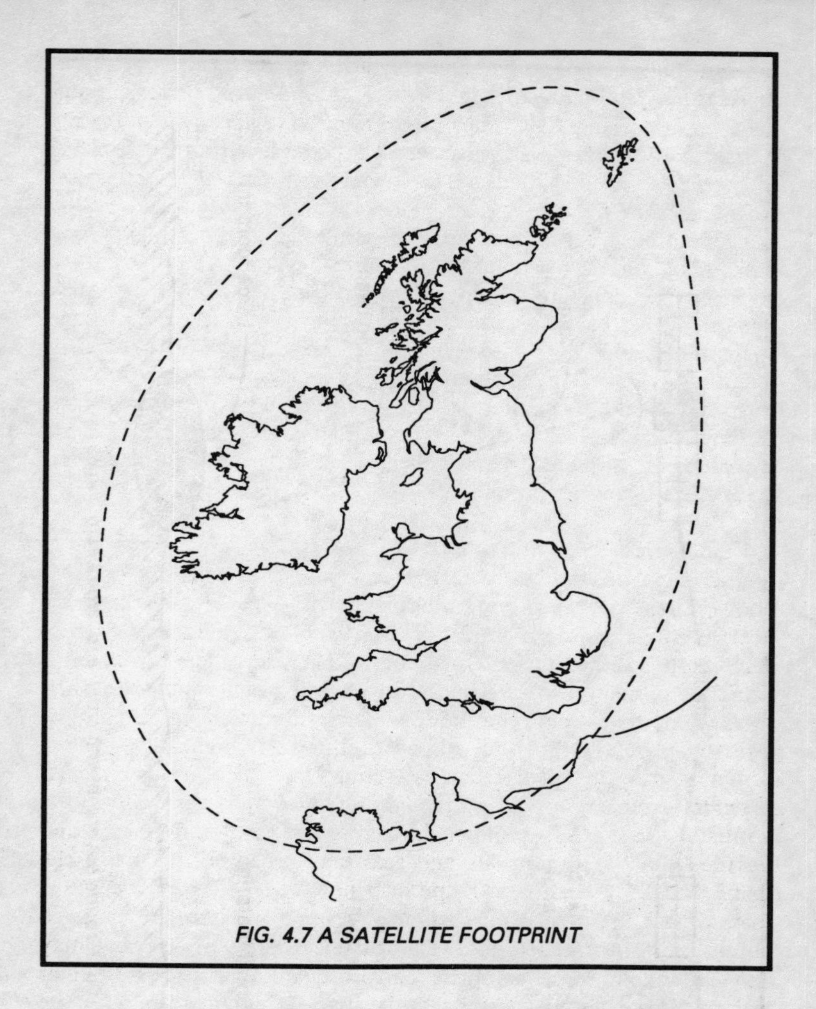

FIG. 4.7 A SATELLITE FOOTPRINT

to grapple with the additional calculations for estimation of the antenna size as also shown in Section 5.2.5. A third method which avoids most of the calculations is to assume a set of standard antenna characteristics such as efficiency, noise performance and losses and then quote on each footprint the minimum antenna diameter which can be used.

4.7 Beamwidth

Figure 4.1 showing how a wave generated at the focus of a parabolic dish is reflected as a narrow parallel beam leaves the impression that

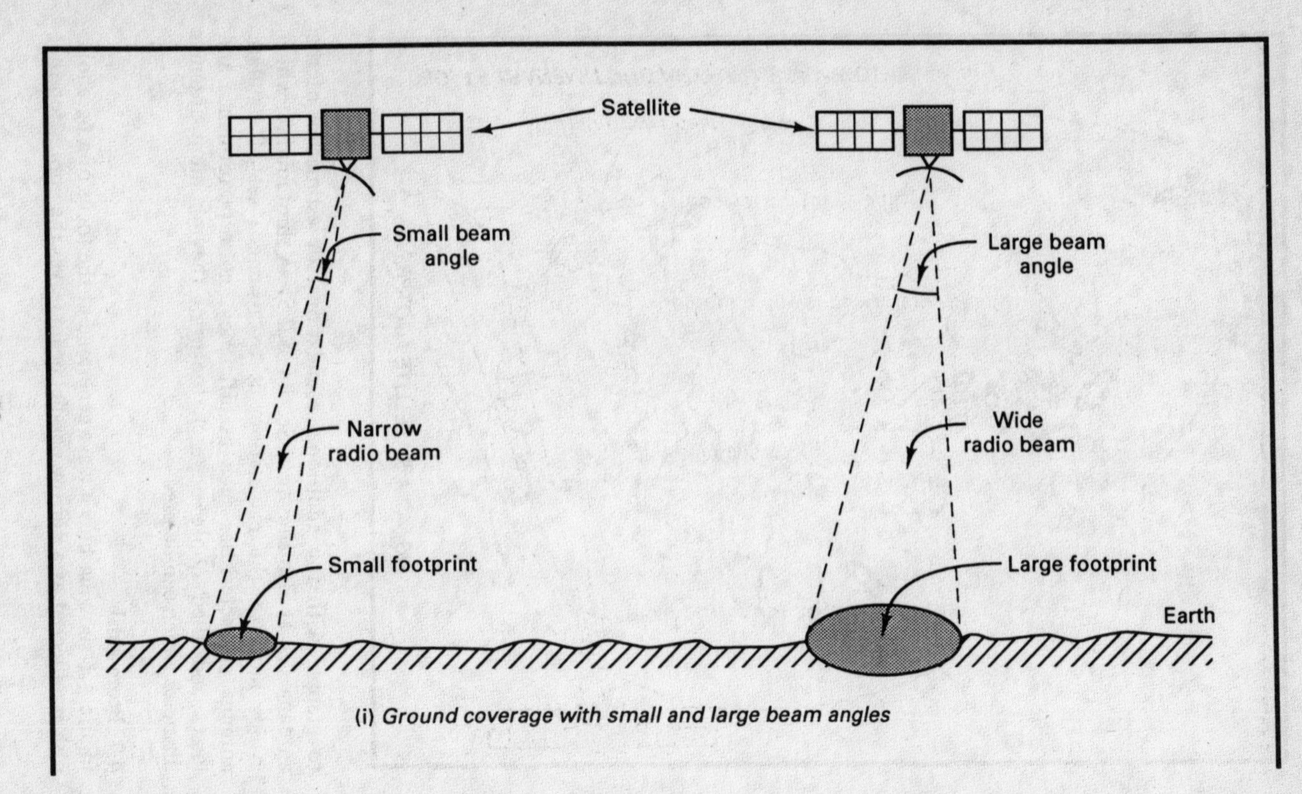

(i) *Ground coverage with small and large beam angles*

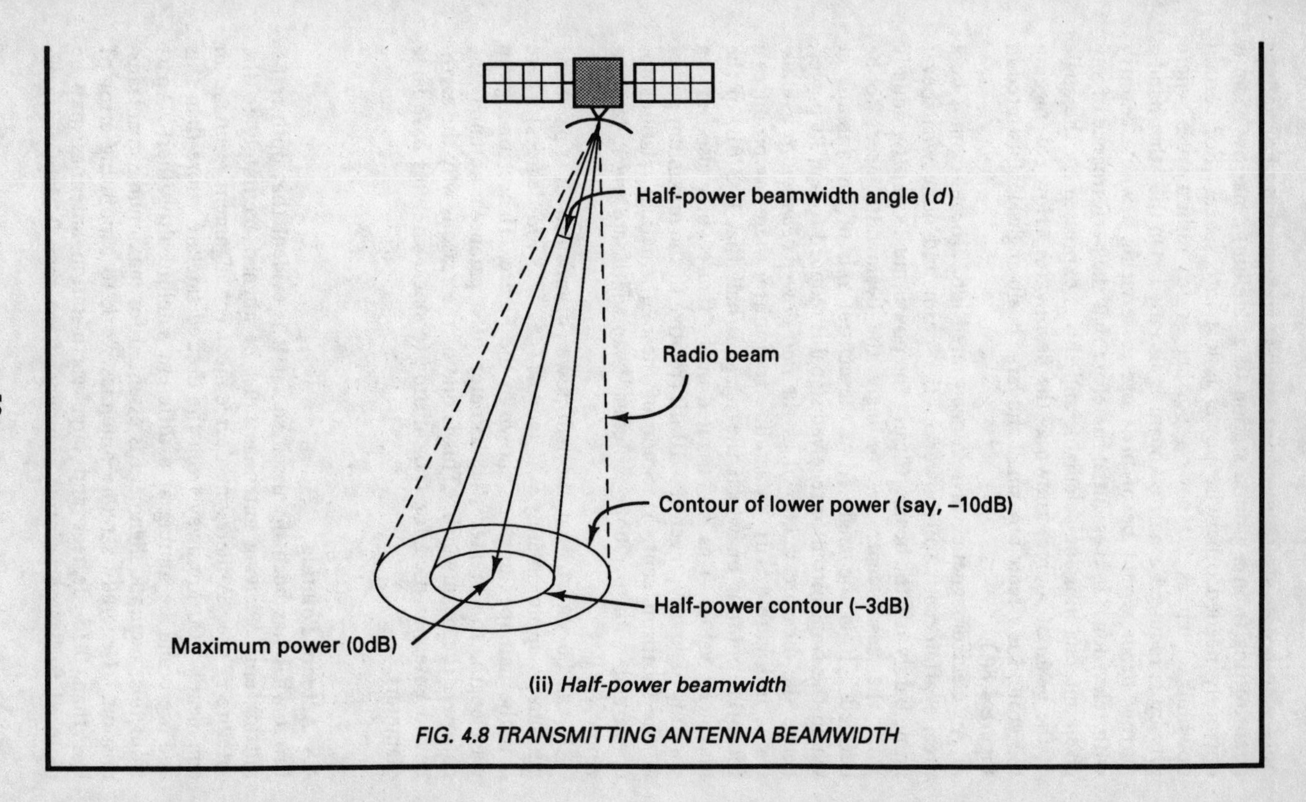

FIG. 4.8 TRANSMITTING ANTENNA BEAMWIDTH

only such beams are required for satellite transmission. For uplinks (transmitting from a ground station to a satellite) a narrow beam is obviously needed otherwise power is lost into space. For certain downlinks such as for news reporting or for military use where privacy is required a narrow beam is also used. On the other hand a narrow beam would be of little use in servicing a whole country with television. It is essential therefore that the performance of any given antenna must be calculable and this is expressed by the angle of the beam at source, known as the *beamwidth.* How coverage on the ground is affected by small and large beam angles is illustrated in Figure 4.8(i).

An electromagnetic wave concentrated into a beam cannot be a precise affair with full power in the beam and none whatsoever immediately off the beam. Clearly the power must tail off gradually from the beam centre, accordingly the beamwidth needs to be defined. It is the angle from the central axis of the antenna at which the transmitted or received signal is reduced to half its power, hence the measure is known as the *half-power beamwidth* as illustrated in Figure 4.8(ii). A satellite transmitter therefore produces a footprint contour on which the signal is half that (–3 dB) of the maximum value. This is illustrated in the figure which also shows a contour of lower power at –10 dB (relative to the maximum power at the footprint centre). Some of the mathematical relationships concerned with parabolic antenna beamwidth are given later in Appendix 9.

It can be shown that the half-power beamwidth of a parabolic antenna is approximately λ/d radians (i.e. 57.3 λ/d degrees) where d is the half-apex angle as shown in the Figure. Hence for a given wavelength, the greater the diameter of a parabolic antenna, the narrower the beamwidth. This is one of the reasons why the earth-bound antennas used for transmitting to single satellites have large diameters.

4.8 Antenna Pointing

From what has been discussed so far it is evident that both transmitting and receiving antennas must be aligned accurately to the satellite in use. A mere one degree incorrect alignment results in an aim nearly 700 km away from the desired satellite, more than this and we could be aiming at a different satellite altogether! Apart from this, accurate alignment is essential for maximum signal transmission. Generally satellite antennas work to a pointing error of less than 0.15 degrees and large earth-station antennas may be required to have a pointing accuracy of less than 0.01 of a degree.

This can only be achieved by automatic systems involving computer control. As an example the *monopulse* tracking system for large receiving antennas employs additional feedhorns fixed round the perimeter of the main dish, each of these produces a radiation pattern slightly different from the main one. If the main dish is pointing accurately the outputs of the perimeter feedhorns balance, if however the main dish is off-beam, there will be higher level signals from one or two of the feedhorns compared with the remainder. Measurement of the differences produces an *error* signal which is used to adjust the position of the main dish so that this signal is reduced to a minimum. Another system of interest turns the antenna a tiny amount in one direction, then measures the change in antenna signal output. If this increases a further turn in the same direction is made and the measurement repeated. This continues until the signal is found to decrease after a move whereupon the dish is moved back. These are not the only earth station tracking systems, several others are in use, each having its own particular advantages.

At the other end of the scale are the portable dishes used by army personnel on the move and news reporters. Such tight tolerances as are used by the large antennas are of course unthinkable in these applications.

4.8.1 Satellite Television Receiving Antennas

The now ubiquitous satellite dish fitted on the outside walls of buildings and in gardens which is equipped with a single low noise block converter (LNB) is so often in view that it needs little physical description. It has a diameter commensurate with the received signal strength. To some these dishes disfigure our homes even more than the existing collections of rods for UHF reception but that is the price we have to pay for a multitude of additional television programmes. Such a dish is illustrated in Figure 4.2(i) or (iii). It must be aligned accurately onto the chosen satellite or motorized so that other satellites can be brought into view as required.

Dish antennas which see one satellite position only are generally known as the *azimuth/elevation* type (Az/El). There may be more than one satellite in the same "position": e.g. ASTRA 1A, 1B and 1C, all at 19.2°E. They are not of course falling over each other, in fact they are many kilometres apart but are so far away that to our dishes on earth, they appear to be transmitting from the same position. Thus to receive any of the channels, no dish adjustment is required so here we can consider the several satellites to be grouped as one only.

Two angles only need to be calculated for any particular location and the dish is set up to them. Of course one can go out or aloft armed with a spanner and change to alternative *Az/El* angles for reception from other satellites but this is hardly recommended for wet and windy days. The alternative is to use a second type of receiving system which employs a *polar mount.* This completely overcomes the adjustment by spanner idea mentioned above. In this case the dish is moved to the appropriate azimuth and elevation angles for a range of satellites by remote control from the comfort of an armchair indoors. Needless to say, it costs more. Further details are given for home installations in Section 7.3.

Siting and Alignment

Firstly there must be an uninterrupted path to the satellite otherwise the oncoming radio wave will be blocked. Even passing aircraft can cause "picture flutter". A dish must therefore be sited so that, except for the inevitable clouds, rain or mist, there is an unobstructed "view" of the satellite not only from the centre of the dish but also from the edge. For many would-be European viewers, the larger type of dish with a clear view from home or garden to around the South presents no problem. On the other hand for others with buildings or trees in the way, things are not so simple with the larger dishes unless a suitable flat roof is available. We must also be conscious of Local Authority regulations, planning permission may be required especially for dishes of more than 90 cm diameter and where the dish is likely to project above the ridge of the roof. Special regulations will probably apply to flats and especially when a house has been converted into flats.

For good reception a dish should be aimed to within about half of one degree so garden dishes, especially the larger ones, must be bolted down onto a solid base or a concrete block. There is no question of a dish being able to move let alone blow over in a strong wind. Dishes of diameter 90 cm or less can be mounted on a wall or on a suitable chimney stack. Fortunately for the majority of viewers the signal strength provided by most television satellites is sufficient for a dish of 60 cm to be adequate for most of the area served with perhaps a dish of 80 cm at the periphery. Pointing a dish in the right direction firstly requires a knowledge of the appropriate angles of azimuth and elevation for the particular satellite. Elevation is easily understood by imagining that the satellite can be seen through a telescope. The angle the telescope makes with the horizontal (as indicated by a spirit level) is the elevation. That the elevation varies with latitude is demonstrated by Figure 4.9(i) where it is seen to be 90 degrees at the equator (satellite directly overhead),

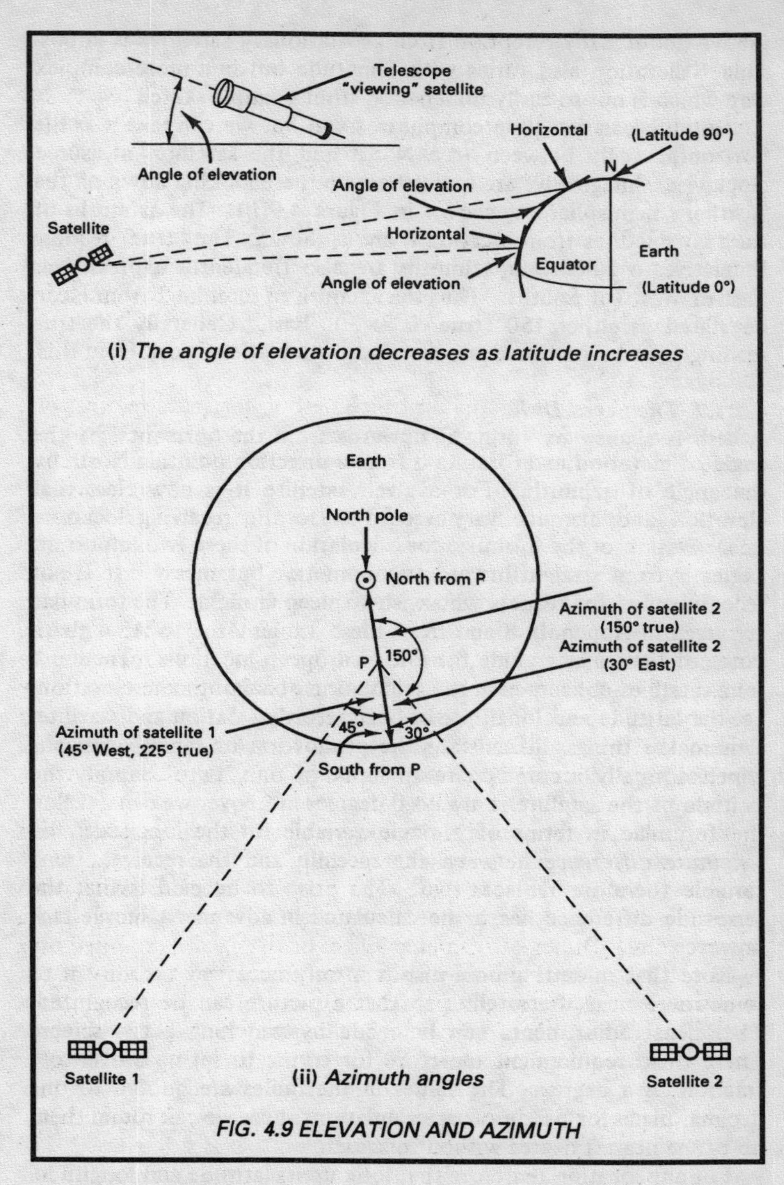

FIG. 4.9 ELEVATION AND AZIMUTH

decreasing as latitude increases. Ultimately reception becomes difficult because the radio beam arrives at such a low angle that it is obstructed by hills, buildings or even trees. At even higher latitudes (above about 81°) reception from geostationary satellites is impossible. Elevation also varies with longitude but in a more complex way which is not so easily understood from a simple sketch.

Azimuth can get quite complicated too but we can take it as the horizontal angle between true North and the satellite, measured clockwise. Imagine we are somewhere up there looking down on the Northern hemisphere as shown in Figure 4.9(ii). The azimuths of the two satellites from location *P* are as shown. The "true" reading is relative to North but azimuths are also frequently expressed as East or West (of South). Thus the azimuth of satellite 2 from *P* can be stated as either 150° true or as 30° East. Generally the true reading is to be preferred because most calculations work from this.

4.8.1.1 The Fixed Dish

A dish is aligned by tilting it upwards from the horizontal by the angle of elevation and rotating it from a direction pointing North by the angle of azimuth. For a given satellite it is now clear that elevation and azimuth vary according to the receiving location.

Derivation of the formulae for calculation of these two important angles is from straightforward trigonometry, but messy. It is not recommended for readers who wish to sleep at night. The formulae are given in Appendix 8 and from these Tables A8.1 to A8.4 give a range of figures as a guide for most European locations. There are four variables concerned in the calculation of azimuth and elevation, i.e. the latitude and longitude of both receiving station and satellite. This makes things difficult for the production of tables or graphs which normally accommodate an input of only two. Happily the latitude of the satellite is always 0 degrees moreover we can develop the formulae in terms of a single variable for the longitude, the *longitude difference* between the satellite and the receiver. One variable therefore replaces two. The price to be paid is that the longitude difference has to be calculated in advance, a simple task however.

Note that in setting up a dish it is only necessary to point it to somewhere near the satellite so that a picture can be recognized. Then final adjustments can be made by watching a t.v. screen. There is no requirement therefore for trying to set up a dish to a fraction of a degree. The values in the tables are quoted to one decimal place for use in other calculations, here we can round them up to the nearest degree without disquiet.

For any location therefore first look up its latitude and longitude. For cities and major towns the figures are normally quoted in a good atlas, otherwise they may be estimated from a map. The longitude of the satellite is also required, this is always available from satellite and allied magazines. We work in decimal fractions of degrees but values quoted in atlases are usually in minutes. These are changed to decimal by dividing by 60 or alternatively if a calculator is not to hand and our arithmetic has seen better days, Appendix 4 can be used.

The practical use of the *Az* and *El* figures in setting up a dish for use in a home satellite receiving system is discussed in Section 7.3.1.1.

We are dealing with angles and use a representation of them by Greek letters, θ (theta) for latitude and ϕ (phi) for longitude with subscripts r for receiving station, s for satellite and d for difference. The difference in longitude, ϕ_d between satellite and earth station is therefore:

$$\phi_d = \phi_s - \phi_r .$$

A complication is that the formulae require that longitude values to the east of 0° are considered negative. Several examples follow to show how it is done. Accordingly we can quickly get a good idea to within a degree or so of the azimuth (*Az*) and elevation (*El*) angles for any place in most of Europe. For those more daring, Equations A8(2) and A8(3) enable calculations to be made for anywhere in the World.

Consider first Figure 4.10 (which is an extension of Figure 4.9) for a better idea of the azimuth calculation. Two receiving stations R1 and R2 are shown. Suppose it is required to calculate the azimuth angle at which the dish has to be set at R1 to receive satellite 1. Then:

$$\text{latitude of R1} = \theta_r = 45^\circ$$

$$\text{longitude of R1} = \phi_r = 45^\circ$$

$$\text{longitude of satellite 1} = \phi_s = 15^\circ$$

hence, longitude difference, ϕ_d, is equal to:

$$\phi_s - \phi_r = 15 - 45 = -30^\circ .$$

The appropriate table in Appendix 8 is A8.2 and at 45° latitude, with ϕ_d = –30 is the value 140.8°. This is the required azimuth angle which is measured from the line joining R1 to the N. Pole.

The elevation to which the dish should be set is to be found in the same place but in Table A8.4, i.e. 30.3°.

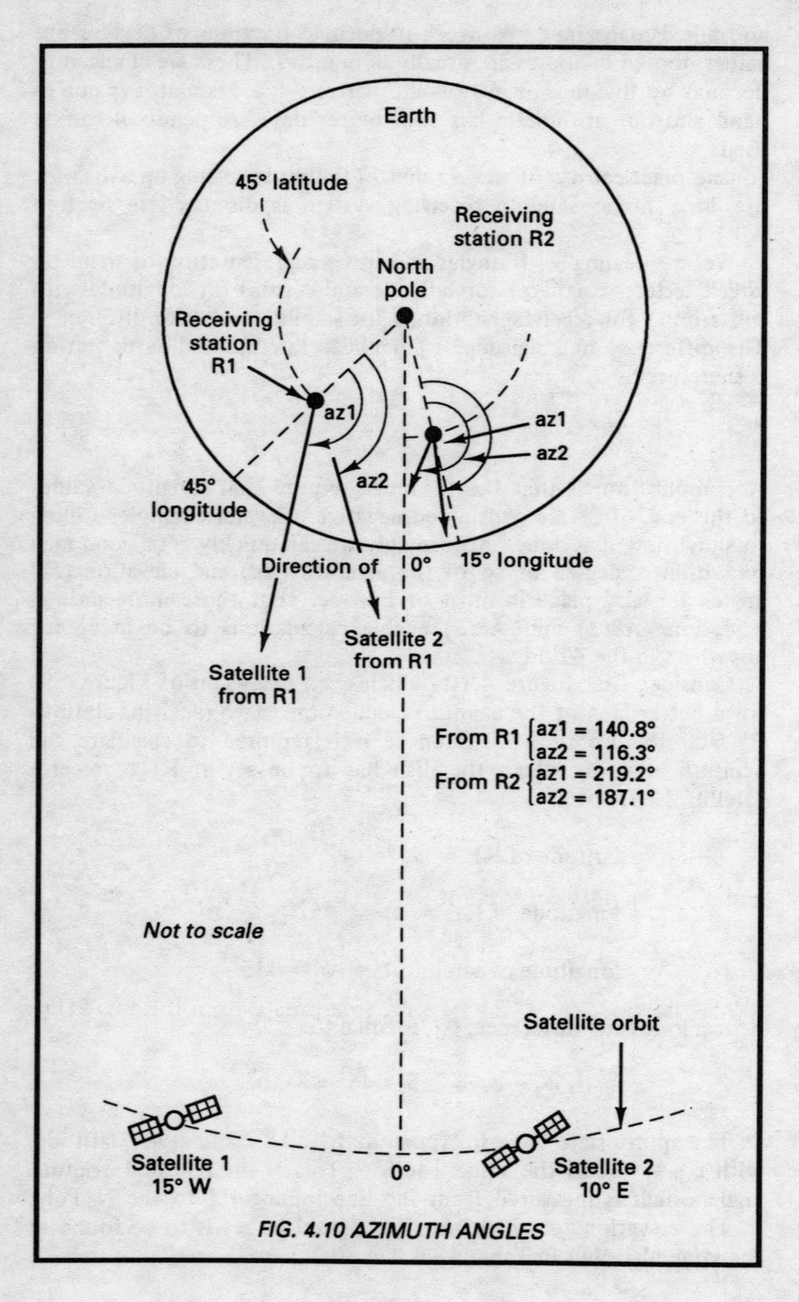

FIG. 4.10 AZIMUTH ANGLES

Next suppose it is desired to realign the dish at R1 to receive satellite 2.

$$\phi_r = 45° \qquad \phi_s = -10°$$

(remembering that ϕ_s is east of 0° and therefore negative)

$$\therefore \qquad \phi_d = \phi_s - \phi_r = -10 - 45 = -55°$$

and Table A8.2 gives for latitude 45°, ϕ_d −55°, 116.3° for *Az* and Table A8.4 gives for latitude 45°, ϕ_d −55°, 15.5° for *El.*

Azimuth angles for the two satellites from location R2 are also given in Figure 4.10. However for more realistic practice let us consider a few actual locations and satellites, the latter being two of the forerunners, INTELSAT VI F1 at 27.5°W and EUTELSAT II F1 at 13°E.

LONDON:

$$\theta_r = 51°.32' \qquad \phi_r = 0°.5'W$$

Changing to decimal (Appendix 4):

$$\theta_r = 51.5° \qquad \phi_r = 0.1°$$

To receive INTELSAT VI F1:

$$\phi_s = 27.5°$$

Then: $$\phi_d = \phi_s - \phi_r = 27.5 - 0.1 = 27.4°$$

From Tables A8.1 and A8.3 for $\theta_r = 51$ and $\phi_d = 27.4$:

$$\underline{Az = 214° \text{ (true)}} \qquad \underline{El = 26°}$$

(Note that for $\phi_d = 27.4$ we have to estimate from the two values for 25 and 30.)

To receive EUTELSAT II F1:

$$\phi_s = -13° \qquad \therefore \phi_d = -13 - 0.1 = -13.1$$

From Tables A8.2 and A8.4,

$$\underline{Az = 164°} \qquad \underline{El = 30°}$$

AMSTERDAM:

$$\theta_r = 52°.22' = 52.4°$$

$$\phi_r = 4°.53'E = -4.9°$$

To receive INTELSAT VI F1:

$$\phi_d = 27.5 - -4.9 = 32.4$$

(remembering that minus minus = plus). From Tables A8.1 and A8.3,

$$\underline{Az = 219°} \qquad \underline{El = 23°}$$

To receive EUTELSAT II F1:

$$\phi_d = -13 - -4.9 = -8.1$$

From Tables A8.2 and A8.4:

$$\underline{Az = 170°} \qquad \underline{El = 30°}$$

MADRID:

$$\theta_r = 40°.24' = 40.4°$$

$$\phi_r = 3°.41' = 3.7°$$

To receive INTELSAT VI F1:

$$\phi_d = 27.5 - 3.7 = 23.8$$

From Tables A8.1 and A8.3:

$$\underline{Az = 214°} \qquad \underline{El = 37°}$$

To receive EUTELSAT II F1:

$$\phi_d = -13 - 3.7 = -16.7$$

From Tables A8.2 and A8.4:

$$\underline{Az = 155°} \qquad \underline{El = 40°}.$$

The very last estimation for elevation may have caused some misgivings for we have θ_r at 40.4 and ϕ_d at −16.7, both inconvenient "in between" values in Table A8.4. If something better than a wild guess is required, the problem is resolved by first interpolating for $\theta_r = 40$, with the result, say 40.6, then for $\theta_r = 41$, say 39.6. Clearly about half way between these two, the result for θ_r is 40.4 which, when rounded to the nearest whole number, is 40. More sophisticated methods can of course be used, even to plotting on graph paper or use of formula but here we are looking for a guide only, not a precise answer.

We conclude this section with the thought that already those readers who are unsure that the requirement of an uninterrupted view of the satellite can be met, are now able to reach a decision. All that is required is that they know where North is and are able to estimate angles, with the help if necessary of a compass and a protractor.

4.8.1.2 The Motorized Dish

So far we have discussed fixed dishes, i.e. those which "see" one satellite position only and most television viewers will be quite satisfied with one of these for many different programmes can be received from a single or group of satellites. The more adventurous however may prefer a motorized dish or *polar mount* because although it can be rotated by hand for alignment with alternative satellites it is more likely to be controlled remotely. In fact with the more sophisticated receivers all that the user needs to do is to key-in the channel required. The system does the rest, it moves the dish, sets the polarity to vertical or horizontal as required and fine-tunes everything.

Compared with the basic *Az/El* mount therefore it has much to offer, not in setting up because this is more difficult but certainly in subsequent use. Obviously the system is more expensive moreover if the user wishes to pick up the more distant satellites then an appropriate size of dish must be used otherwise the received signal will be too weak. As an example, to receive programmes intended for the U.K. in say, Greece or Cyprus would require a dish of at least 6 metres in diameter. To understand how it all works we need first to appreciate what might be referred to as the *heavenly arc.*

When or before we get up in the morning the sun looks at us *from the horizon* in the east. At midday it is high in the sky (the meridian) and finally it disappears *at the horizon* in the west. Were we to plot its position regularly throughout the day, it would be found that the path is in the form of an arc. The sun is in fact following a *polar curve* which is defined as one related in a particular way to a given curve and to a fixed point called a *pole* (nothing to do with the earth's poles). Hence to follow a polar curve, not only must its shape be known but also the position of the pole. With a little imagination it is possible to see that from any earthly location the part of the geostationary orbit which contains available satellites also follows a polar curve.

Such a curve can be produced mathematically but we have an easier way out by using data from Tables A8.1 – A8.4 of Appendix 8. Take a couple of locations as an example, London at say 51° latitude, Barcelona at 41°.

Start with Table A8.1 at latitude 51°. For $\phi_d = 0$ the azimuth is 180°. A satellite in this position must be due south and at the highest point, the pole. The elevation is found from Table A8.3 and for latitude 51° and $\phi_d = 0$, this is 31.6°. Azimuth and elevation may then be plotted as for example in Figure 4.11. Plotting for other values of ϕ_d (at 51° latitude) generates the right hand half of the London curve as shown. The left part of the curve arises from Tables A8.2 and A8.4 at 51° (remember ϕ_d is negative for easterly directions). The two halves of the curve are mirror images. The curve for Barcelona is also shown in the Figure with elevations higher than those for London as would be expected. To add a little realism some satellites have been added in their present positions. An interesting exercise for it indicates the order of things pictorially. The arc is actually crowded with satellites but not all broadcast television signals.

The arc is followed by a dish when the latter has a *polar mount*. When fitted with such a mount the dish is attached to its support by a pivot inclined in such a way to the horizontal plane that when turned, a polar curve is automatically followed. The idea is not new, it has been used with telescopes for many years. Accordingly, for anywhere on earth, rotating the dish in the horizontal plane either manually or by motor, sweeps through the arc so that the dish picks up each satellite in turn. Details on setting up a polar mount are given in Section 7.3.1.4.

Returning to Figure 4.11 we note that a satellite (the Intelsat VB F15) is shown well to the left at around 60°E on the London arc. This particular satellite at present broadcasts to countries well to the east (Germany, Turkey, etc.). It has been added to demonstrate that the arc has its limits for here the elevation is only of the order of 10°, hence apart from the low received signal strength requiring a large dish, at such a low elevation there might be the problem of earthly objects getting in the way.

Just one of the many motorized dish arrangements is shown in Figure 4.12. This particular type is operated by remote control

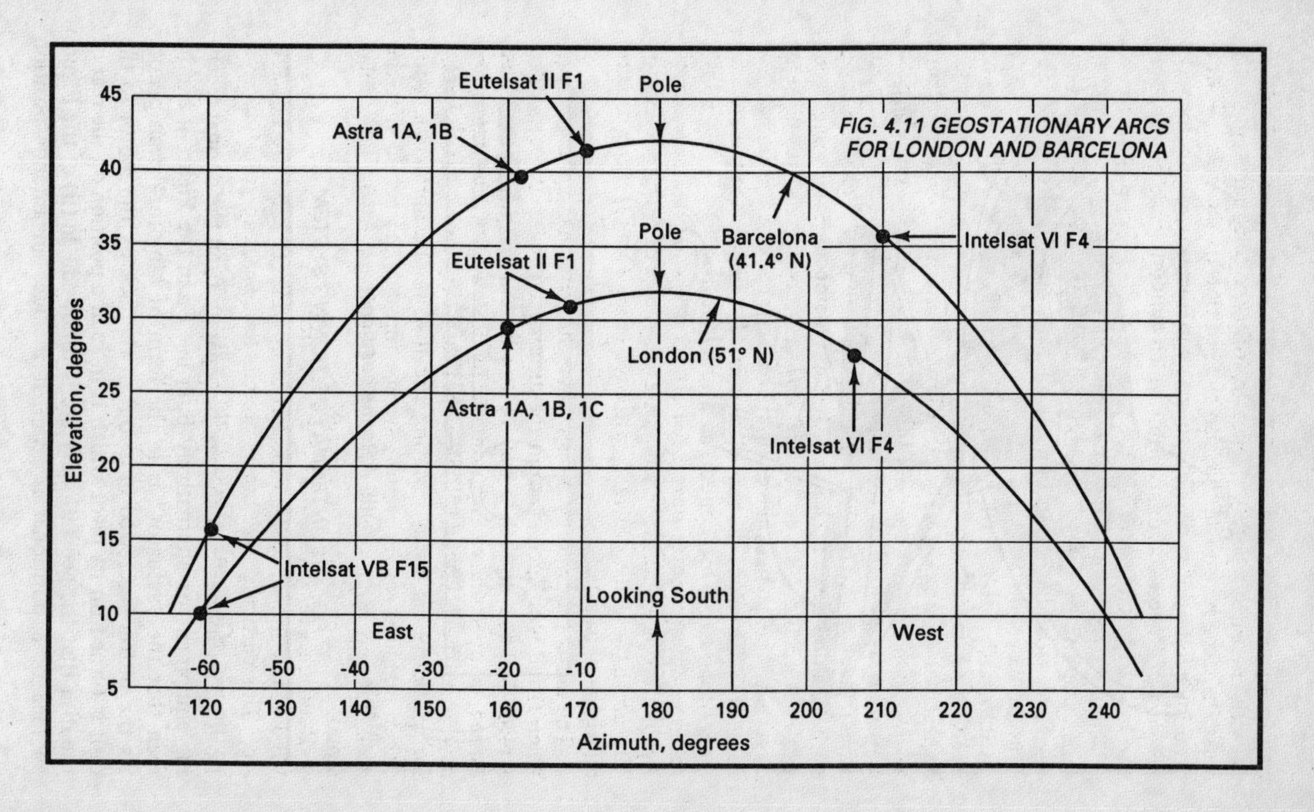

FIG. 4.11 GEOSTATIONARY ARCS FOR LONDON AND BARCELONA

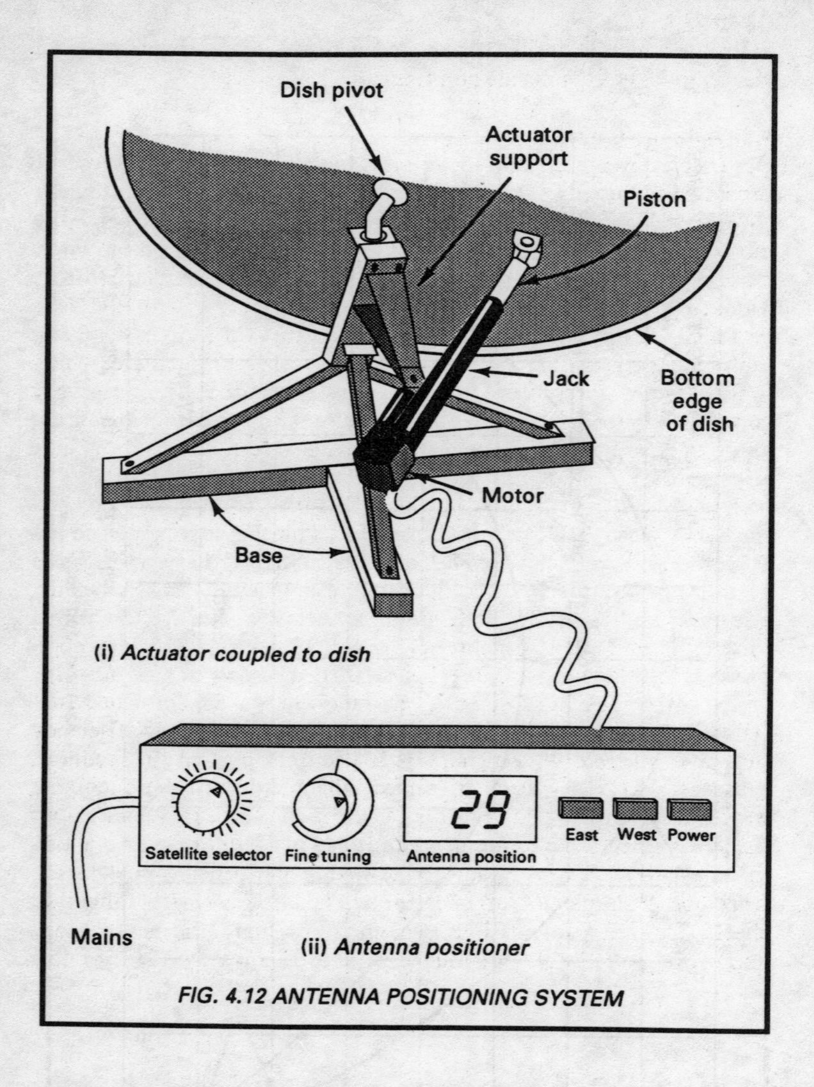

(i) *Actuator coupled to dish*

(ii) *Antenna positioner*

FIG. 4.12 ANTENNA POSITIONING SYSTEM

from a "positioner" which may be separate from the normal receiving equipment or alternatively built-in. From the Figure it can be seen that the "actuator" is in the form of a motor-driven piston. As the piston is screwed in or out of the jack by the motor so the dish is turned on its pivot which is in such a position on the rear of the dish that the geostationary arc is followed. In (ii) of the Figure an idea of a positioner is given, it includes an antenna position indicator which is useful, otherwise on switching on the user has no idea as to where the dish is pointing.

4.9 Antenna Losses

Ground receiving antenna losses are generally greater than those of the satellite because the latter are kept under constant accurate control. Such control is necessary because both the sun and the moon and even other planets exert gravitational forces on satellites, especially the geostationary ones which are nearer. Such forces produce a slight drift in the satellite position in a figure-of-eight pattern each day. Ground control ensures that this drift is kept to a minimum, this is part of the *station-keeping* programme which unfortunately consumes fuel. Large ground stations employ computer-controlled tracking systems so that pointing losses are kept to a minimum and hence the signal-to-noise ratio is not impaired.

Small receiving antennas on the other hand are subject to several small but nevertheless important losses. Pointing loss is mentioned in Section 4.8.1., this arises because it is unlikely that the dish will be pointing *exactly* towards the satellite even with fixed dishes for television and especially with mobile systems as used by the armed forces. Prime focus antennas [Fig.4.2(i)] are less efficient compared with the offset type [Fig.4.2(iii)] because of the *blocking effect* of the feedhorn system. In fact when a well-designed subreflector system is in use [Fig.4.2(vi)] the overall antenna efficiency can be as much as 20% higher. It is also unlikely that the feedhorn system (the LNB) will be in perfect alignment with the oncoming wave. With time dust and dirt enter the scene and experience shows that even insects sheltering in some types of feedhorn create a loss. In addition there are the inevitable wiring and other copper losses. It is not a simple job to measure these losses so with tongue in cheek, for later calculations we might estimate for an ordinary home or commercial parabolic dish a total antenna + wiring loss (L_T) of 4.5 decibels.

Chapter 5
TRANSMISSION

Section 1.5 points out how important it is that any electromagnetic signal carrying information is recognizable above the general noise level of the system. For that reason it is important that in all transmission aspects an overall signal-to-noise ratio appreciably exceeding 1 is obtained. Very clever computer programs exist which are capable of extracting signals when the signal-to-noise ratio is poor with some degree of success, their use however is mainly restricted to the security services. In this chapter we must always be aware of the restrictions noise imposes on any transmission channel. We can sum this up by the golden rule that for a satisfactory overall received signal not only must the signal be of sufficient amplitude but also the signal-to-noise ratio must be adequate.

5.1 Satellite Distance

Clearly the distance between a satellite and a particular location on earth depends on the latitudes and longitudes of both. When the latitude and longitude differences are both zero, then the satellite is directly overhead and its distance away (d) is equal to h, the vertical height. Clearly the distance increases with the latitude of the earth station and it also increases with the difference in longitude between the satellite and the earth station.

In Section 4.8 it is shown that the satellite latitude is 0° and that the longitude considerations can be simplified by using the longitude difference, ϕ_d. The receiving location we designate θ_r. Then the distance (d) from the satellite to a transmitting or receiving terminal can be shown to be:

$$d = h\sqrt{\{1 + 0.42(1 - \cos\theta_r \cdot \cos\phi_d)\}}$$

where h is the height of a satellite above the equator (35,786 km).

As an example, the distance to a satellite at 27.5°W from Liverpool at 53.4°N, 3.1°W is calculated from:

$$\phi_d = \phi_s - \phi_r = 27.5 - 3.1 = 24.4° .$$

Hence:

$$d = 35786\sqrt{1 + 0.42(1 - \cos 53.4 \cdot \cos 24.4)}$$

$$= 35786\sqrt{1 + 0.42(1 - 0.5962 \times 0.9985)}$$

$$\therefore d = 39{,}070 \text{ metres.}$$

5.2 Basic Transmission Equations

A signal arriving on earth is weak because of the losses sustained on its long journey. We develop first a basic transmission equation which expresses the received signal power available at the output terminals of a receiving parabolic antenna but assuming that the path from the satellite transmitter to the ground receiver is via free space only. All these considerations have to be in terms of the isotropic antenna which as we have seen in Section 4.3 is perhaps the most convenient reference antenna for this particular purpose. We are obliged to use this comparison technique because the performance of most practical antennas cannot be assessed directly.

Let the transmitting antenna gain (relative to the isotropic) $= G_t$

Let the receiving antenna gain (relative to the isotropic) $= G_r$

Power supplied to the transmitting antenna $= P_t$

Then the radiated power flux density (p.f.d.) at a distance d metres:

$$\text{p.f.d.} = \frac{G_t P_t}{4\pi d^2} = \frac{\text{eirp}}{4\pi d^2} \qquad (1)$$

where eirp is the effective isotropically radiated power (the eirp enables us to rate the gain of a practical antenna relative to the isotropic in a particular direction).

Let A_{eff} be the effective area of the antenna where $A_{eff} = \eta A$, A being the physical area of the antenna and η its efficiency (Sect. 4.4). Then, receiving antenna gain:

$$G_r = \frac{4\pi A_{eff}}{\lambda^2} \qquad \text{i.e. } A_{eff} = G_r \times \frac{\lambda^2}{4\pi}$$

(as developed in Sect.4.4). From this:

$$P_r = \text{p.f.d.} \times A_{eff} = \frac{G_t P_t A_{eff}}{4\pi d^2} \qquad (2)$$

where P_r is the available power at the terminals of the receiving antenna, hence:

$$P_r = \frac{G_t P_t G_r}{4\pi d^2} \times \frac{\lambda^2}{4\pi} = G_t P_t G_r \left[\frac{\lambda}{4\pi d}\right]^2 \qquad (3)$$

This is the basic transmission equation in which free space path loss, L_{fs} is equal to:

$$\frac{P_t}{P_r} = \left(\frac{4\pi d}{\lambda}\right)^2 \qquad (d \text{ and } \lambda \text{ in metres})$$

or in decibels:

$$L_{fs} = 20 \log 4\pi d - 20 \log \lambda \ \text{dB} \qquad (4)$$

or expressed in terms of frequency in GHz and d in kilometres:

$$L_{fs} = 92.44 + 20(\log f + \log d) \ \text{dB} \qquad (5)$$

A reminder – this is for free space, there will be additional losses due to the earth's atmosphere so a correction for this is considered below. Summing up this Section, we find that free space path loss is a measure of the signal attenuation between two isotropic antennas separated by a distance, d in free space. The isotropic antennas are themselves loss-free by definition. What we must not lose sight of here is that L_{fs} does not indicate that power is absorbed as the wave travels through space for there is nothing in space to do this. It simply accounts for the fact that from an isotropic antenna the electromagnetic wave spreads out in all directions, most of which is lost. Put simply, the free space loss is a measure of the power which is lost into empty space.

5.2.1 Atmospheric Loss

The free space path loss does not tell the whole story for it assumes clear space and clean air. There is an additional loss to take into account due to the earth's atmosphere and appropriately known as *atmospheric loss.*

The boundary between the atmosphere and space is not clearly defined and the effect of the various "layers" can only be expressed by a single loss value as found by measurement. There is the analogy of light from the sun which manages to get through even when dark heavy clouds are around but with less brightness. Radio frequencies are also affected by diffusion and absorption in the lower layers of the atmosphere and by rain, mist and clouds.

The total atmospheric loss, L_{at} obviously increases with the length of the path through the atmosphere. The path length is related to the angle of elevation, i.e. the loss is greater for the lower elevations. A rough guide is given by:

elevations of 5 – 14°, L_{at} = 5.0 dB

elevations of 15 – 24°, L_{at} = 2.5 dB

elevations of 25 – 45°, L_{at} = 1.7 dB .

As with the weather there can be nothing precise about this. There is only a small chance that L_{at} will rise to greater values and even then by usually not more than a few decibels. Good design must therefore include an allowance for L_{at}.

5.2.2 Power Flux Density on Earth

From Section 5.2:

$$\text{p.f.d.} = \frac{\text{eirp}}{4\pi d^2} \quad \text{(where eirp } = G_t P_t\text{)}$$

Changing to decibel notation and including the atmospheric loss, L_{at}:

$$\text{p.f.d.} = \text{eirp} - 71 - 20 \log d - L_{at} \quad \text{dBW/m}^2 \qquad (6)$$

where d is in kilometres, eirp in dBW and L_{at} is in dB.

5.2.3 Receiving Parabolic Antenna Output

From Section 5.2:

$$P_r = \text{p.f.d.} \times A_{eff} = \text{p.f.d.} \times \eta A$$

(see also Sect.4.4), where A is the physical area of the antenna (m^2) equal to $\pi D^2/4$ where D is the diameter in metres and η is the antenna efficiency as a fraction. Accordingly:

$$A_{eff} = \frac{\pi D^2 \eta}{4} \; .$$

Let the total receiving antenna losses (i.e. including wiring loss – see Section 4.9) equal L_r. Then:

$$P_r = \text{p.f.d. (dBW/m}^2) + 10 \log \left(\frac{\pi D^2 \eta}{4}\right)\text{(dB)} - L_r \text{ (dB) dBW} \quad (7)$$

5.2.4 Carrier-to-Noise Ratio

So far we have considered the effects of noise in terms of the signal-to-noise ratio (s/n). The signal however is usually a variable quantity especially with speech and music as shown by the voltage output of a microphone. Noise may be easily measurable but on the other hand it is more likely to be sporadic. The difficulties in measuring and quoting a single signal-to-noise ratio for any particular set of circumstances are therefore evident. There are of course complicated equipments and procedures to minimize this problem, e.g. continuous measurements of the level of the signal and/or noise averaged over one or more seconds, but such methods are mainly confined to the laboratory. How the various noises combine in a typical downlink is illustrated in Figure 5.1.

In radio transmission we are fortunate in that the radio signal is a carrier, modulated in some way by speech, music, data, etc. and generally the modulation is such that overall the mean level of the transmission is reasonably constant and in this case it is the level by which the carrier exceeds the noise which is important. Accordingly when we are concerned with transmission by radio it is usually more convenient to consider the problem of noise in terms of the *carrier - to - noise ratio* (c/n) and this is generally how it is handled in satellite transmission. From this it is clear that any satellite link must have a sufficiently high c/n ratio for successful operation.

The effect of clouds is small compared with that of the more dense raindrops, cloud attenuation not being appreciable until well over 30 GHz whereas rain attenuation becomes effective above about 5 GHz. The sun itself is also a strong but variable source of noise but which is unlikely to create interference with a wanted signal for more than say, 10 minutes on a few days in any one year.

Fortunately satellites use line-of-sight transmission in directions nearly perpendicular to the atmosphere. Link fades arise when the signal is reduced momentarily through atmospheric attenuation and for satellite transmission both the frequency and duration of link fades are therefore lower than for terrestrial microwave transmission. The effects of man-made noise are also less. In fact most noise on a satellite link is generated in the receiving system.

Because a satellite receiving system consists of many different units (amplifiers, converters, demodulators, etc.), an assessment of the *equivalent system noise temperature* is made to avoid the complexity of combining the effects of the various and differing noise contributions (see Appendix 5). T_s stands for this temperature and the way in which we arrive at a single value for T_s is illustrated in Figure A5.1.

From Equation (3) and Appendix A5.1.2:

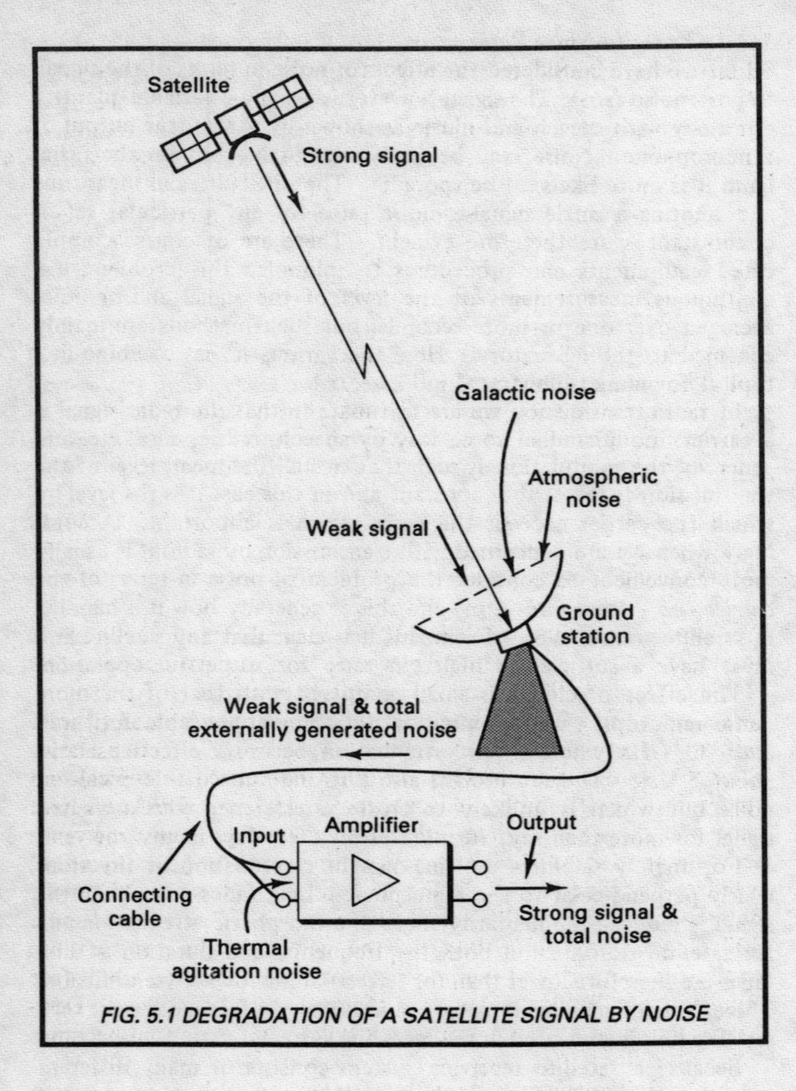

FIG. 5.1 DEGRADATION OF A SATELLITE SIGNAL BY NOISE

$$\mathrm{c/n} = G_t P_t G_r \left[\frac{\lambda}{4\pi d}\right] \times \frac{1}{kT_s B} \tag{8}$$

or in brief:

$$\mathrm{c/n} = \frac{P_r}{kT_s B}$$

where k is Boltzmann's constant and B is the bandwidth (usually that between the upper and lower 3 dB points) and as a reminder, T_s is the temperature to which a resistance connected to the input of a noise-free receiver would have to be raised in order to produce the same output noise power as the original receiver. It is therefore a convenient measure of noise.

All the terms in Equation 8 may be considered as constant except for G_r and T_s, hence:

$$c/n \propto G_r/T_s .$$

This is generally known as the G/T ratio (gain over temperature) which is widely used as a figure of merit in assessing the overall performance of a receiving system as far as noise is concerned. If we consider a satellite receiver together with its antenna, it can be assessed on the overall performance typically as follows:

antenna gain (relative to the isotropic - Sec.4.4)	39 dB
system noise temperature	25 dB
hence system figure of merit (G/T ratio)	14 dB

(see Appendix 2 for decibels). This particular figure of merit is extremely useful in determining overall system performance when the signal level on the ground is known for the satellite.

As a very simplified example of the use of the c/n ratio, a noise power budget is developed below. This is only an illustrative example, many of the values can only be calssed as typical and therefore do not apply to any particular situation. Designers have access to data from which estimates of antenna equivalent noise temperature due to water and oxygen in the atmosphere, sky noise, etc. can be made (e.g. T_a in Fig.A5.1 at an elevation of 25° might be around 50 K). An LNB might have a noise temperature of some 300 K. Next suppose that for the whole receiving system the equivalent system noise temperature is calculated at, say 400 K (see Appendix 5). The noise power is then calculated as follows. Decibels again make life easier for which Boltzmann's constant becomes –228.6 dBW/K/Hz.

From Boltzmann's basic relationship between energy and temperature, the available noise power, P_n generated by thermal agitation of electrons is given by:

$$P_n = kTB \text{ watts} .$$

The bandwidth, B is generally taken as the IF bandwidth for the IF amplifiers act as a band-pass filter on the whole system. Then:

B at 27 MHz	= 74.3 dB relative to 1 Hz
T_s at 400 K	= 26 dB relative to 1 K
k	= −228.6 dBW/K/Hz .

Hence:

$$P_n = -128.3 \text{ dBW}$$

and we see that this is the total noise power calculated via the system noise temperature. Knowing the power level of any particular carrier therefore the c/n ratio can be calculated.

This has become rather technical so we might usefully consider some practical results. For television a satisfactory carrier-to-noise ratio is essential otherwise the picture will suffer from spots and in the extreme break up altogether. The extreme occurs when the c/n ratio is around 6 dB or less for then the carrier is not greatly in excess of the noise which if in any way sporadic, can really produce a messy result. At the other end of the scale, a c/n of 20 dB is more or less perfection but it must be admitted, not many systems are as good as this. Somewhere in between, say well above 6 dB but not necessarily as high as 20 dB is the practical requirement and Astra transmitting to a 60 cm antenna well within the footprint gives us an example for at a c/n of 14 dB the picture is generally assessed as good and in no way inferior to a terrestrial television picture. Carrier-to-noise figures for satellite systems other than television of course vary according to the use and tolerance of the particular system.

5.2.5 Link Budgets

It is now appropriate for us to combine the various transmission considerations of this Section into a *link budget* so that we can estimate the signal output of a receiving antenna, knowing of course the transmitted power of the satellite. In this exercise we are not restricted to considering parabolic receiving antennas only, horn or flat antennas are included provided that the essential characteristics are known. A link budget simply adds up the gains and losses a signal experiences on its way down from the satellite amplifier to the output of a ground receiving antenna. This is the downlink, we need not consider the uplink in the same way because for any particular satellite it is generally a single link carefully designed so that the satellite output is at the required power level and at an adequate carrier-to-noise ratio. The basis of all our calculations is the isotropic antenna, a purely theoretical component yet as Section

4.3 shows, we cannot get away from it. We will find that most noise on such a downlink is generated not so much in the link itself as in the receiving equipment on the ground.

A link budget is most conveniently expressed in decibels for then analysis is reduced simply to addition and subtraction. Such budgets are applicable to any satellite system but here we take as an example just one of the many television distribution satellites.

Table 5.1 gives the budget details, helpful notes follow with references to more information as required.

Table 5.1 – DOWNLINK SIGNAL POWERS

Satellite - ASTRA 1A.

Receiving Location - LONDON (51.3°N, 0.1°W)

SATELLITE:		
	Position	19.2°E
(1)	ϕ_d	–19.3°
(2)	Frequency	11.406 GHz
(3)	Transponder output power	45 watts (+ 16.5 dBW)
(4)	Antenna gain (G_t)	34.5 dB
(5)	Waveguide and wiring losses (L_{wt})	1.0 dB
(6)	e.i.r.p.	50 dBW
TRANSMISSION PATH:		
(7)	Length (d)	38744 km
(8)	Basic path loss (L_{fs})	205.3 dB
(9)	Atmospheric loss (L_{at})	1.7 dB
(10)	Total path loss	207 dB
(11)	P.f.d. on ground	–114.5 dBW/m²
RECEIVING ANTENNA:		
(12)	Diameter (D)	80 cm
(13)	Efficiency (η)	65%
(14)	Gain (G_r)	37.7 dB
(15)	Losses (L_r)	4.5 dB
(16)	A_{eff}	–4.9 dB
(17)	Output signal power	–123.9 dBW

Notes:

(1) See Section 4.8.1.1. ϕ_d is the difference in longitude between the satellite and the earth station.

(2) This is the frequency of a single television channel. The range

for this particular satellite at present is 11.214 – 11.436 GHz.
(6) See Section 4.3.1.
(7) See Section 5.1.
(8) From Equations (4) or (5) of Section 5.2.
(9) See Section 5.2.1.
(10) Simply the addition of the basic path loss and the atmospheric loss.
(11) From Equation (6) of Section 5.2.2.
(14) From Table A6.1 (Appendix 6).
(15) See Section 4.9.
(16) From Table A7.1 (Appendix 7).
(17) Here we note that a measure of the output of a receiving antenna may be given by:
  (i) raising the input signal level by the gain, G_r, here calculated from e.i.r.p. (6) – total path loss (10) + receiving antenna gain (14) – antenna losses (15); or
  (ii) multiplying the surrounding power flux density by the effective area of the antenna, i.e.: p.f.d. on ground (11) + A_{eff} (16) – antenna losses (15).

It may be found that some footprint maps are labelled according to the satellite e.i.r.p. This is not so informative because they quote what the satellite transmits rather than what arrives on earth. However change from e.i.r.p. to p.f.d. is simply carried out as above using Equation (6) of Section 5.2.2.

It is also important that the approximations are kept in mind:

(i) the atmospheric loss is an average figure, it could be less for much of the time but on the other hand it is liable to increase substantially for very short periods;
(ii) the dish receiving loss has been estimated at 4.5 dB. We could be lucky in having the dish and LNB aligned exactly, in which case the allowance is excessive.

It is all very well developing a link budget as above but we end up with a figure which does not seem to mean much. Evidently we have yet to decide whether this signal level is satisfactory. Naturally it all depends on the purpose of the system, e.g. television, military, weather data, etc. In this section it is television which interests us most and we must accept that it is not a high power, high quality service which is generally required but one which is acceptable to most viewers who require the simplest and cheapest of installations for receiving the "home" satellite. Little is left in hand for the inevitable rainy day and there may be some increase in sky noise because of the large beamwidth of a small dish (Sect.4.7). The one we have chosen for demonstration happens to be a satellite broadcasting to the UK and known to give rise to a generally satisfactory television picture so we can assume that the *link margin* is ample.

The link margin is of utmost importance in satellite engineering. It is a figure which expresses the degree by which the carrier power exceeds the minimum allowable value. Link margins vary with the system use and although to some extent can be calculated in advance, in practice they are usually determined by decreasing the satellite output power until the received signal is sufficiently degraded to be unacceptable. The difference between this signal and its working value is the link margin. Generally satellite link margins ensure that operation is satisfactory for at least 99% of their operating time, sometimes as high as 99.9%.

Table 5.1 considers a satellite distributing television programmes to Europe including many channels to the UK. As such its output power is known to be sufficient. However we happen to have chosen an 80 cm diameter parabolic receiving antenna of 65% efficiency. Other dish sizes and efficiencies can easily be substituted in the Receiving Antenna section of Table 5.1. Note that there is little point in searching satellite magazines for the efficiency of particular dishes, it is seldom published, however the manufacturer or agent will have the information.

Using the technique of Table 5.1 we can move on to assess the possibilities of satisfactory reception from any satellite. We assume that the p.f.d. on the ground is quoted on the foot print contour maps (Sect.4.6), if not then it can be calculated as for the transmission path in Table 5.1. Note that this table is concerned with a comparatively small area only (London). Larger areas are contained within footprints at power levels which are lower than the maximum, here we take a figure of 3 dB lower.

Appendix 7 gives us help in Table A7.1. Any practical dish may be chosen for the basic calculations then the table quickly assesses the effect of changing to dishes of alternative dimensions. We have seen that once the p.f.d. on the ground is known, the dish output signal power follows from:

$$\text{p.f.d. (dBW/m}^2\text{)} + A_{\text{eff}}\text{ (dB)} - \text{total losses (dB)}$$

from which it is evident that the dish output relates directly to changes in A_{eff} since the p.f.d. and total losses remain constant. If the dish is not circular, A_{eff} is calculated from 10 log (ηA). As an example, with any satellite, changing from a 60 cm, 50% efficiency to a 120 cm, 65% efficiency dish improves reception by $(-1.3 - -8.5) = 7.2$ dB.

The technique is especially useful when attempting to assess the prospects of tuning in to satellites not directly intended for us. Take for example a far away satellite for which the p.f.d. on the ground is quoted as –120 dBW/m^2. But first let us be sure of our standard. On the assumption that the dish output signal from Astra 1A is satisfactory within its service area, then an output signal power of at least (–123.9 – 3) = –126.9 dBW is required. Assume losses as before (line 15) of 4.5 dB. Then, using Table A7.1:

	dish diameter (cm)	*efficiency* (%)	A_{eff} (dB)	*output signal power* (dBW)
(1)	180	65	+2.2	–122.3
(2)	120	60	–1.7	–126.2
(3)	90	55	–4.6	–129.1
(4)	60	55	–8.1	–132.6

Dish:

(1) excellent because it produces an output signal 4.6 dB better than required;

(2) good because it produces an output signal power at a level known to be satisfactory;

(3) not so good – the output signal power is 2.2 dB down;

(4) doubtful, some 5.7 dB down. This might result in a grainy picture and even loss of frame or line hold at times.

These few calculations already point the way. If the particular satellite is only to be used occasionally, then of course the smaller dishes may be acceptable. The prospects with satellites further afield can similarly be assessed.

Finally to gain more confidence in the assessment method, consider the master antennas for cable subscribers. These need to ensure a good picture through highly unfavourable weather conditions, even from the less powerful satellites. The antennas are of 3 – 4 metre or even greater diameter. Take for example a 4 m dish, efficiency 60%, working to a satellite which gives the UK a p.f.d. on the ground of –118 dBW/m^2. From Equation (7) of Section 5.2.3 and allowing only 3 dB for losses (more accurate alignment), the dish signal output is –112.2 dBW, ample for a reliable service.

5.3 Frequency Bands

Section 1.4.2 shows that satellites generally need to work at frequencies in excess of some 1 GHz so that electromagnetic waves can travel straight through the ionosphere with little or no loss. A second reason for the use of such high frequencies is that the higher the carrier frequency the greater is the bandwidth of the modulation we can impress on it, e.g. up to 500 MHz or even more. As an example, over a 500 MHz system, by using both vertical and horizontal polarizations together (Sect.4.5) more than 20 × 36 MHz channels can be provided. The frequency range of such a channel is more than adequate for transmission of a television signal, many thousands of telephone circuits or digital signals at many millions of bits per second.

It will be seen that for most satellites the uplink is usually provided on a higher frequency band than the downlink. Why a frequency change from uplink to downlink is necessary is explained in detail in Section 5.6 which shows that with no such change, the whole system becomes unstable. Now Section 4.4 notes that the gain of an antenna is proportional to the square of the frequency. Section 4.7 and Appendix 9 consider beamwidth of antennas and although not specifically stated, it can be deduced that the gain of an antenna is also inversely proportional to its beamwidth, which on second thoughts looks reasonable enough. In a nutshell therefore, as frequency rises, so does the antenna gain but its beamwidth decreases. Hence using the higher frequencies on a downlink would entail more problems of alignment of the receiving antennas because the satellite itself moves daily about its mean position. Accordingly the lower frequency band is chosen for the downlink. For the uplink, at the higher frequencies the increased antenna gain results in an improved signal-to-noise ratio.

A further reason for the choice is that the efficiency of a satellite power amplifier tends to decrease as frequency increases. Since d.c. power for operation of the amplifier is at a premium up aloft it is therefore expedient that the downlink frequency should be lower than that of the uplink for which there is a plentiful supply of power to the transmitter.

The earlier non-military satellite systems used a range of frequencies generally designated as frequency band 4/6 GHz where 4 GHz is somewhere in the middle of the downlink range (3.7 – 4.2 GHz) and 6 GHz is within the uplink range (5.925 – 6.425 GHz). These bands lie within the frequency range known as the C band. (The classifications "C band" and the "Ku band" which follows have been adopted generally – they were earlier allocated for military and radar purposes, note however that the band frequency ranges quoted by different authorities vary.) The C band was first developed for commercial satellite communication, it has the advantage of low sensitivity to man-made and sky noise.

A second frequency band is now being used extensively as equipment technology advances, it is designated 11/14 GHz with a downlink range 10.95 – 11.7 GHz and uplink, 14 – 14.5 GHz. This is within the Ku band. These frequencies are sufficiently high that they are not shared with terrestrial microwave services as are some frequencies in the C band. Higher satellite downlink powers are therefore permitted although higher rain attenuation may be a problem.

A third frequency band is reserved for military use. It is designated 7/8 GHz, actually running from 7.25 to 8.4 GHz.

5.4 Modulation Systems

As we have seen, a radio wave can be projected into the air, atmosphere or space by a transmitting antenna. It can travel a short or very great distance and finally be picked up by a receiving antenna. A single frequency wave doing this however carries no information whatsoever except that it is "on" or "off". To carry speech, television or data (i.e. *information*) it must be changed in some way according to that information. The ways of doing this are known as *modulation* and this is in fact the basis of all radio transmission. Basically a high frequency wave is made to carry the lower frequency information, the higher frequency being known appropriately as the *carrier* and what it carries is known as the *baseband*. For technical reasons the frequency of the carrier wave must be many times that of the baseband frequency it transports. Terrestrial television provides an example where a carrier frequency of some 200 MHz or more carries a baseband (the television signal) of about 6 MHz.

A modulator is a complex electronic circuit for which the symbol only is shown in Figure 5.2. The baseband input (the modulating frequencies) and the carrier frequency are fed into the modulator. These are then mixed in a special way and the modulated wave is taken from the output terminals as shown. Imagine if you can a 200 kHz carrier wave and consider that it has to carry (i.e. be modulated by) a 500 Hz continuous piccolo note (about upper C on the musical scale). The simplest method by which this can be achieved is known as *amplitude modulation* (a.m.), the method generally used for medium wave broadcasting. Unfortunately for us it is not used for satellite transmission so we have to struggle with a more complicated method known as *frequency modulation* (the well-known "f.m."). With this the carrier wave has constant amplitude but its frequency varies. The *degree* of frequency variation is proportional to the amplitude of the modulating wave whereas the *rate* of variation is according to that of the modulating wave itself [see Equation (1) of Appendix 10]. Thus the 200 kHz carrier wave

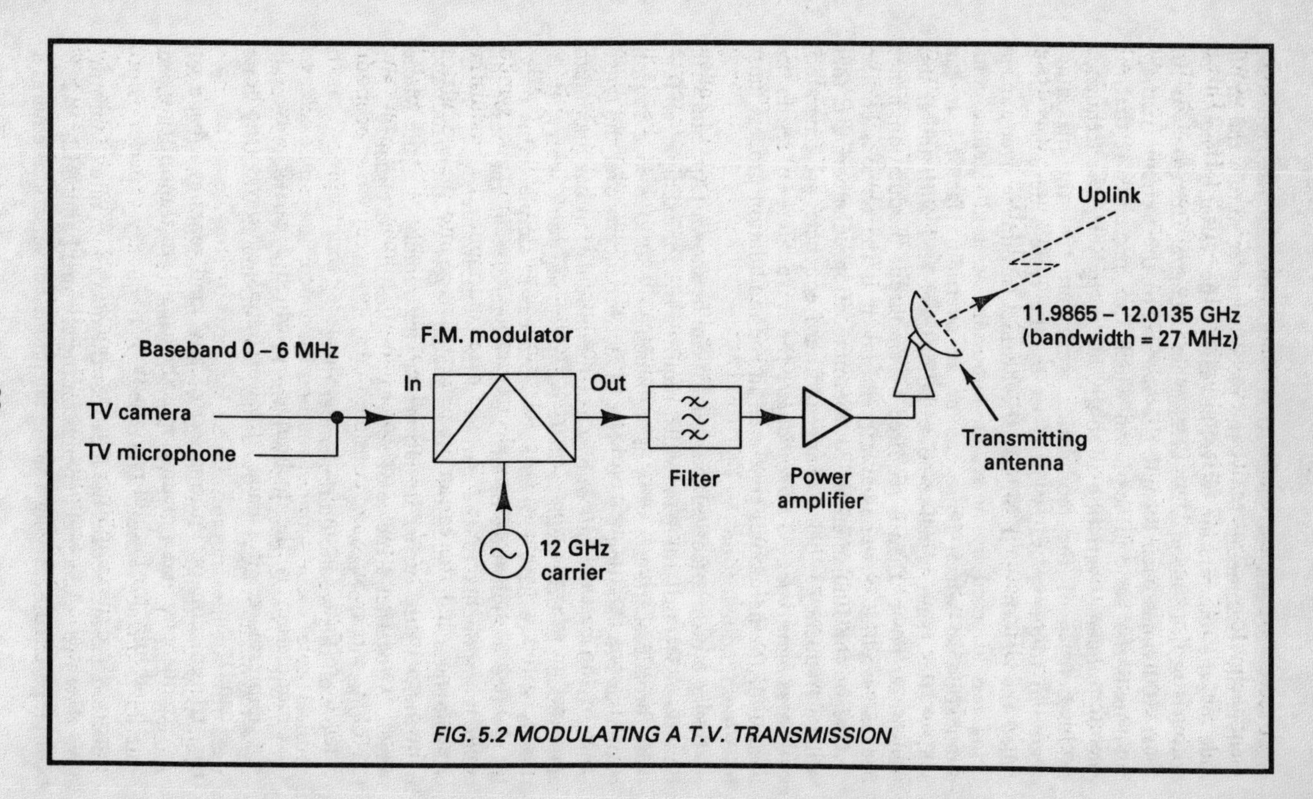

FIG. 5.2 MODULATING A T.V. TRANSMISSION

might vary between 190 and 210 kHz 500 times each second when modulated. If someone blows the piccolo harder so that the amplitude of the modulating wave increases, the variation could be between say, 180 and 220 kHz, again changing from one to the other 500 times each second. Simplified to the extreme perhaps but this is the basic principle on which f.m. rests. Note that the *rate* of frequency variation is controlled entirely by the frequency or frequencies of the modulating wave but that the maximum *degree* of frequency variation is set by the system designer. It is called the *deviation.* F.M. has the distinct advantage over a.m. in being less affected by noise, an important asset in satellite communication in which noise can be an overriding factor. It can achieve the same signal-to-noise ratio (Sect.1.5) with much less transmitter power than is required by amplitude modulation, hence because satellite power is at a premium, it is the obvious choice.

One of the fundamentals of communication engineering is that while a frequency is being changed, other frequencies are generated, known as *harmonics.* To *demodulate* the wave later, i.e. regain the baseband, the harmonics generated in the modulation process must be present, none lost on the way.

Taking a more practical example, it can be shown [Appendix 10, Equation (3)] that the bandwidth required by, for example, an f.m. radio broadcast transmission is as much as 180 kHz so a station quoted as broadcasting on 90 MHz in fact broadcasts over the range 89.91 to 90.09 MHz but only when the modulation has maximum amplitude. If this seems a little complicated, don't worry, it is, that is why we have looked at the process in simple terms only. In practice a multitude of different frequencies in the baseband create components of the carrier shifting in frequency at indescribable rates. In addition the changing levels of the bandband frequencies cause carrier deviations varying rapidly over the whole range. Altogether a mêlée of activity, impossible to visualize. But f.m. works and works well.

From all this we should remember that:

(i) modulation is the technique by which a carrier wave has impressed on it a band of lower frequency waves (the baseband);
(ii) transmission of information of any kind requires a band of frequencies and a channel must be capable of transmitting this band fully for faithful reproduction.

Section 1.5 introduces the idea that the maximum rate of information flow over a channel depends on the channel bandwidth. As an example, a circuit carrying a single telephone conversation requires a bandwidth of less than 4 kHz. A single satellite channel is capable of catering for many megahertz. Accordingly telephony systems occupying satellite channels employ modulation techniques to build up high capacity units consisting of many thousands of separate telephone circuits for which might be required a total of 60 MHz bandwidth. Digital systems also transmit enormous quantities of information on systems of large bandwidth (see Section 5.5). Appendix 10 develops the mathematics leading to an idea of the bandwidth required for a frequency modulated wave.

We can now appreciate that for convenience, only the centre-channel frequency is quoted in satellite literature, the bandwidth depending on the type of transmission. As an example, current television transmissions require a bandwidth of some 27 MHz as shown in the Appendix. This is developed in Figure 5.2 which shows a 6 MHz baseband modulating a 12 GHz carrier. An adjacent channel must therefore have a carrier frequency at least 27 MHz different from the one shown. In fact, if we consider for example the satellite Astra 1A, it will be found that the vertically (or horizontally) polarized channels have either 29 or 30 MHz spacings. Note that every link in the chain of uplink, satellite and downlink must be capable of transmitting at least this band, if one link only imposes a restriction, then the whole channel is restricted.

5.5 Digital Signal Processing

Digital transmission is introduced in Section 1.6. It is a means of conveying information over any distance, ending up with a signal relatively undistorted by noise and other channel imperfections. Digital services over satellites range from those which are digital all the way (e.g. computer to computer) to those which are analogue in origin but are converted to digital for ease of transmission (e.g. a telephony circuit). Noise on the way is of course a problem and can cause errors in the received signal but this is minimized by suitable error control techniques (see below).

There are three basic methods of modulating a carrier by binary pulses as shown in Figure 5.3. Note that the number of carrier cycles shown in each case is ridiculously low but must be drawn so for the purpose of illustration. Also here *keying* is the general term used for making or breaking an electrical circuit – as in telegraphy.

(i) *Amplitude-shift keying* (ASK). This is the simplest method, the carrier is present for a digital 1, with no carrier for a 0. This looks simple enough but in fact keying systems require a certain bandwidth for the pulses to be transmitted accurately. This is because anything which is done to a sine wave (the carrier) such as distorting the wave or even switching it on or off generates harmonics.

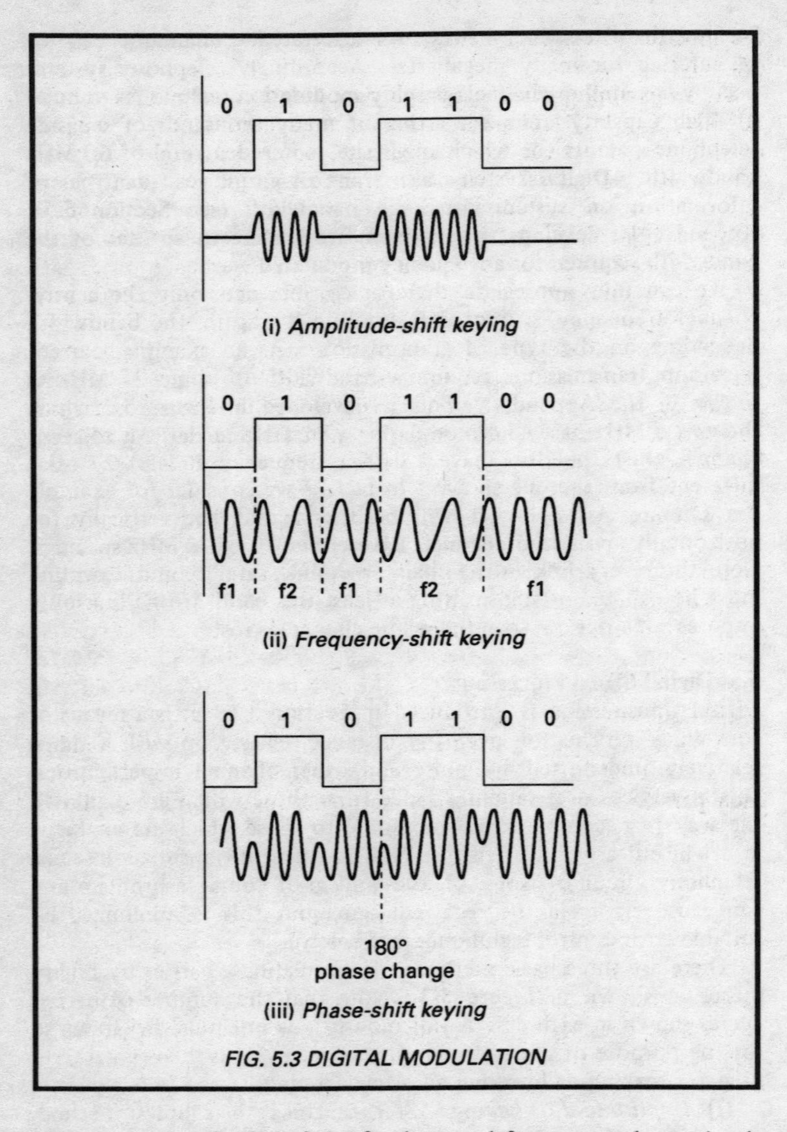

FIG. 5.3 DIGITAL MODULATION

These are at multiples of the fundamental frequency, decreasing in amplitude as the frequency increases. In this particular case, if the maximum baseband signalling frequency is designed by f_S (the maximum occurs when the pulses alternate regularly between 1 and 0) and the carrier on which the pulses are impressed by f_C, then the modulated carrier extends over a frequency range $(f_c - f_s)$ to $(f_c + f_s)$.

As an example, if the baseband signalling frequency (f_s) were at 10 MHz (0.01 GHz), modulating a 4 GHz carrier (f_c), the transmission would extend over the range, 3.99 to 4.01 GHz. This is a completely theoretical example for in practical digital systems the rate of information flow is expressed by the *bit rate* (the number of binary digits transmitted per second) not the frequency as we have shown. The relationship between bit rate and frequency varies with the system and the method of keying in a rather complex way.

(ii) *Frequency-shift keying* (FSK). Again see Figure 5.3 which shows an example. The carrier is switched rapidly between the two frequencies which represent 1 and 0. In the sketch the frequency is lower for a 1 than for a 0. FSK requires twice the bandwidth of ASK because two different frequencies are changed but it is less affected by noise. It has a lower probability of error compared with ASK.

(iii) *Phase-shift keying* (PSK). This is sketched in (iii) of the figure. It will be seen that as the binary signal changes from 0 to 1 or from 1 to 0, the carrier voltage reverses its direction, technically it is said to undergo a 180° phase change. As with ASK it is a single carrier frequency system and it can be shown that a PSK system requires the narrowest bandwidth compared with ASK and FSK for the same rate of data flow. PSK also has the lowest probability of error of all three systems.

A more advanced form of PSK is known as QPSK, the Q standing for *quadrature.* This handles two binary digits at once and allocates each pair to a carrier phase advancement, for example as follows:

45° = binary 00
135° = binary 01
225° = binary 11
315° = binary 10

hence each state of the carrier contains two bits of information and also caters for all binary numbers. It can be shown that QPSK requires only half the bandwidth needed by a PSK system. The equipment required is of course more complex.

5.5.1 Error Correction

Theoretically to be transmitted unscathed a pulse needs fast rise and decay times which implies infinite bandwidth because any restriction of harmonics rounds of and spreads a square wave. In the limit of course, with no harmonics transmitted at all, a pulse emerges as a sine wave. The practical bandwidth therefore is that which ensures that any two adjacent pulses are clearly distinguishable. However errors cannot be eliminated entirely, be it ever so small, there is always some risk hence system design normally includes *error-checking.* It only requires a single pulse representing a binary 1 to be interpreted as a 0 (or vice versa) for the whole character to be read as one entirely different. By the use of various error correction and coding systems, reliable communication can be ensured where normally it would not be possible.

A system is therefore rated by its BER, i.e. the *bit error rate*, which is the ratio of the number of bits received in error to the total transmitted. A good system may have a bit error rate not exceeding 1 in 10^6 for most of the time.

Several techniques are in use for the detection of errors, perhaps the simplest to understand is the one known as *parity checking.* Taking the simplest code which requires 7 bits in length to express a single character, there is one bit spare in the normal 8-bit word (known as a *byte*). This eighth position is conveniently used for a *parity* (equality) bit which we might describe as a single bit used solely to indicate that all is well or not.

An *odd parity generator* adds a single logical 1 when doing so makes the total number of 1's in the byte odd. An *even parity generator* similarly arranges for an even number of 1's. Thus as an example the decimal number 78 might be transmitted as:

1 (parity) followed by 1001110 for odd parity

0 (parity) followed by 1001110 for even parity

(the parity bit might equally *follow* the code). At the receiving end every byte is tested for the chosen parity. A positive result in the test is a good indication of successful transmission. The system has some weaknesses but in most situations these are greatly outweighed by the overall reduced likelihood of undetected error.

Much depends on the type of information being carried. A single error in a bit stream representing a telephony signal would pass unnoticed by the recipient, on the other hand the same error in a computer system could be more serious. More complicated systems are therefore available but the general rule applies that when an error correcting system is added to an existing system, maximum information flow is reduced simply because "redundant" symbols have to be added as we have already seen with parity checking above. However the reduction in information flow is more than offset by the decrease in the likelihood of error.

In simple terms therefore the encoder adds redundant symbols for checking purposes, at the other end of the channel the decoder, knowing the encoding technique, checks what is received and is then able to make an estimate of the original information.

A.R.Q. systems (Automatic Transmission Request) are used with data and telegraphy systems, they provide high reliability and the probability of an undetected error getting through is low. Because effectively the system relies on acknowledgement by the receiver of what it sees as a correct signal or sequence, the round-trip delay of a geostationary satellite system tends to extend transmission time and is therefore rather a disadvantage.

5.5.2 Speech Interpolation Systems

When telephone links are costly as when undersea cables and satellites are being used, efforts are made to increase the number of links possible within a given bandwidth. The earliest system was known as Time Assignment Speech Interpolation (TASI), used for undersea circuits crossing the Atlantic. As the name indicates, telephone users are assigned a channel only while they are actually talking, this is possible because generally only one person speaks at a time and even then there are gaps within the speech itself. Of course, switching must be fast in that a talker is allocated a channel within a few milliseconds of speaking. By this means the number of channels a system can carry is approximately doubled. Nevertheless such a system has its imperfections as we see later. TASI was designed as an analogue system throughout but more recently digital speech interpolation (DSI) has been developed and this is especially useful over satellite systems. It has a better transmission performance and overall the equipment is more likely to be less costly.

In any such system there is always the probability that both participants talking over a telephone channel may speak simultaneously so seizing a channel in each direction. It is therefore evident that systems with only a few channels will be less efficient than the larger ones. Even with the larger systems there will be occasions when there are more active talkers than there are satellite channels. When this occurs an additional talker will encounter "freeze-out" and the speech will be lost until an unoccupied channel becomes available. A low percentage freeze-out is therefore essential so that very little of the first part of any utterance is lost (front-end clipping) and the effect will be unnoticeable by the listener.

Briefly, a voice detector at the transmitting end of a system checks each of the incoming terrestrial telephone circuits in turn. A rise in the speech level classes a circuit as active and an outgoing satellite channel is connected. At the same time a signal is transmitted to the distant end to advise which ongoing telephone circuit should be connected to that particular satellite channel. The whole process is carried out so quickly that the telephone users are usually unaware that their conversation is being switched rapidly between many different satellite channels, unless of course traffic flow is sufficiently low that there are ample unoccupied channels available. Although highly complex equipment is required at both ends of a speech interpolation system, the cost is far outweighed by the increased number of channels available.

5.5.3 Multiple Access

In this section we look at just one of the systems used to provide several ground stations with access to a single satellite channel on a time-sharing basis, *Time Division Multiple Access* (TDMA). This is accomplished by each station transmitting its data up to the satellite in short bursts, all of which are interleaved in time. Because the bursts from different stations go up at different times and arrangements are included so that there is no overlapping, all carriers can be at the same frequency. A typical use of the system is by many news-gathering terminals all transmitting to the main publishing headquarters.

Figure 5.4 shows the method pictorially for a small group of stations – A, B and C. A short burst of data from A (by short is meant less than one millisecond) is transmitted to the satellite. A then ceases to transmit while B and then C transmit their own bursts in sequence. This constitutes a single *frame* on completion of which a new one commences. Timing is controlled by one of the transmitting stations which generates a reference burst at the beginning or within each frame. This reference burst keeps the whole system synchronized. Each burst contains not only the data being transmitted but also digits indicating the start of the burst, the orginating and destination addresses, the packet sequence number, error correction and finally the end of the burst.

It can be shown that a TDMA system running at slightly over 100 Mbit/s and with a frame length of some 2 ms can cater for well over 1,000 station channels. Furthermore the number of channels can be approximately doubled by adding a DSI system (Digital Speech Interpolation – see previous section).

This is only one of the many systems developed for digital transmission by satellite. There are many others designed to cater for different requirements, most having the basic aim of maximum information flow per unit of bandwidth with minimum risk of error (see also Section 8.2 and for example, Figure 8.1).

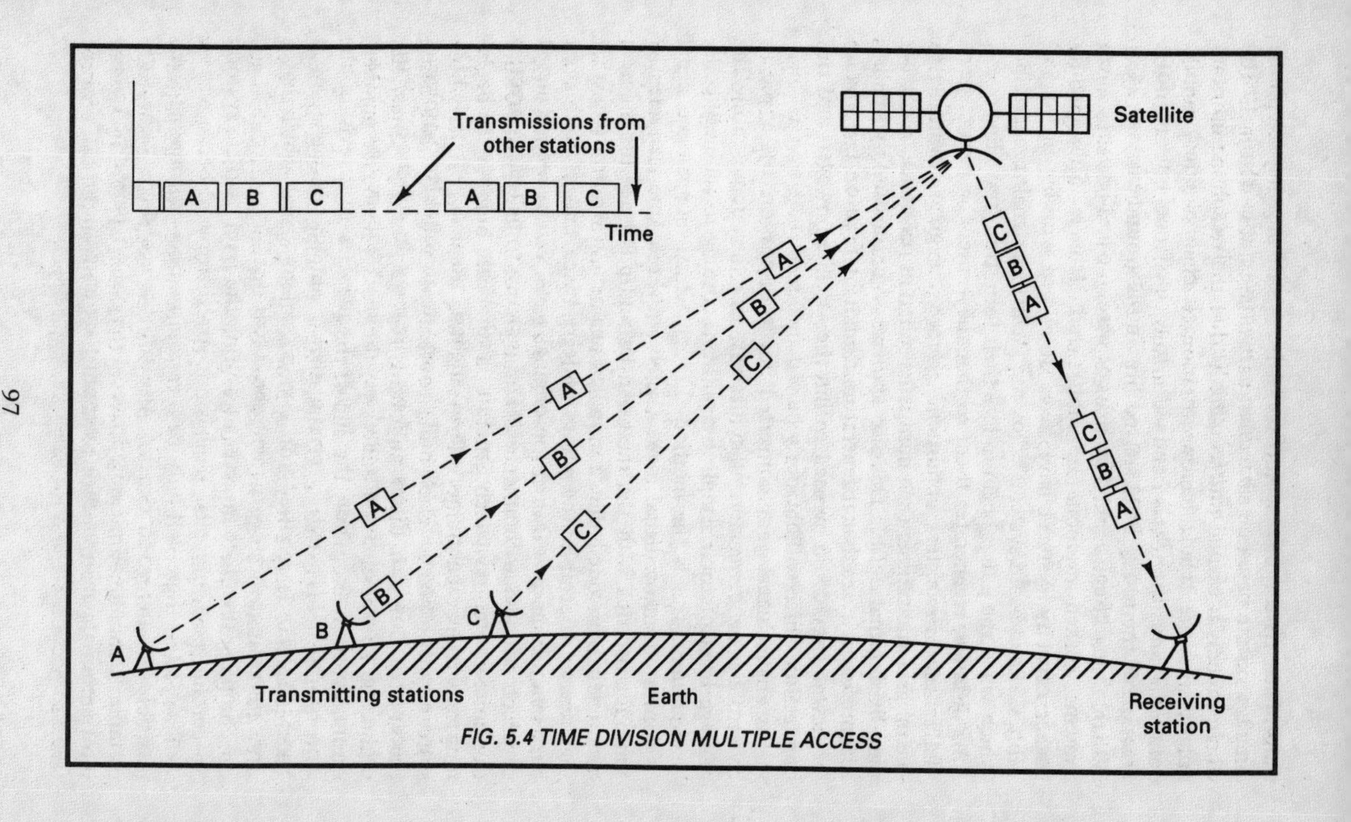

FIG. 5.4 TIME DIVISION MULTIPLE ACCESS

5.5.4 Pulse Code Modulation (PCM)

This is a system based on *time division multiplexing* (t.d.m.). Whereas *frequency division multiplexing* (f.d.m.) allocates a different fraction of the whole bandwidth to each channel continuously, t.d.m. allocates the whole bandwidth to each channel but only for a fraction of the time. PCM started life on the ground carrying 24 or 30 telephony circuits over two pairs of wires (go and return). These systems worked at about 1.5 Mbit/s and 2Mbit/s. Subsequently higher capacity systems have been developed working up to 560 Mbit/s. Land-based systems enjoy the facility of signal regeneration, which when the pulses begin to lose their form from travelling over a long distance, regenerates them so that they enter the next stage of the journey in perfect condition. Hence the great advantage of p.c.m. as a transmission medium is its inherent resistance to corruption by external noise. This also applies to transmission by satellite although it is clear that regenerators cannot be used on the way up or down. They could be used within the satellite however, and this is a facility which will obviously be provided soon.

Analogue signals are converted for transmission by a p.c.m. system by *sampling* at regular intervals and then representing the voltage level of each sample by pulses according to a binary code.

Sampling: this is the method through which an analogue signal is converted into a digital form. H. Nyquist, the American scientist first got to grips with the mathematics and published his theorem stating that provided that a complex analogue waveform (e.g. that of speech) is sampled at a rate at least twice that of its highest frequency component, then the original signal can be reconstructed from the samples without error. This is shown to be true in practice by considering telephony speech carried by a practical p.c.m. system. The speech waveform highest frequency is 3.4 kHz, theoretically this could then be sampled at 6.8 kHz. However because the low-pass filters used for telephony do not cut off sharply, some allowance is made and a sampling frequency of 8 kHz is more appropriate. For the higher frequencies of television the theory still holds good so we find that a sampling frequency of at least 11 MHz is required for a video colour signal which has a maximum frequency component about half this.

Sampling therefore involves measuring the level of the waveform at certain fixed intervals. This is not straightforward because as we will see later, there will only be a certain number of sampling levels possible for any given system. Also to obtain a single value for a sample it should be measured over a very short period of time so that amplitude changes during sampling are insignificant.

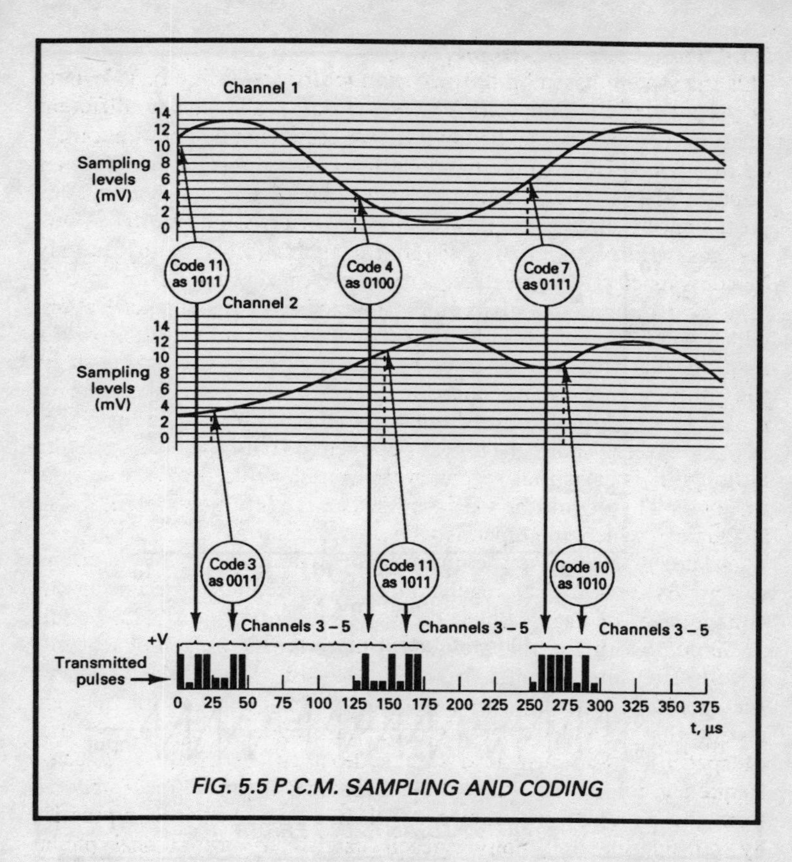

FIG. 5.5 P.C.M. SAMPLING AND CODING

Let us consider a practical system working to 8 bits per sample for which there can be $2^8 = 256$ different codes. Effectively this means that the amplitude range of the signal cannot be sampled precisely but only to the nearest of the 256 levels. This is in fact working in steps and representing a signal by steps or discrete levels in this way is called *quantizing.* To compress 256 levels into a diagram is hardly a practical venture hence Figure 5.5 shows the elements of a less realistic, 4-bit system of only 5 channels. Note that a 4-bit system can only accommodate 16 sampling levels.

The input modulation in the Figure to Channel 1 is a sine wave of 3.4 kHz. It is assumed that Channel 1 is sampled at $t = 0$, 125, 250, . . . , μs (i.e. at 8 kHz). At $t = 0$ the nearest sampling level is at 11 mV. Any value between just over 10.5 and 11.5 mV is interpreted as 11 and the inaccuracy of quantizing begins to show. It is known as the *quantizing error.* The practical system has 256 levels

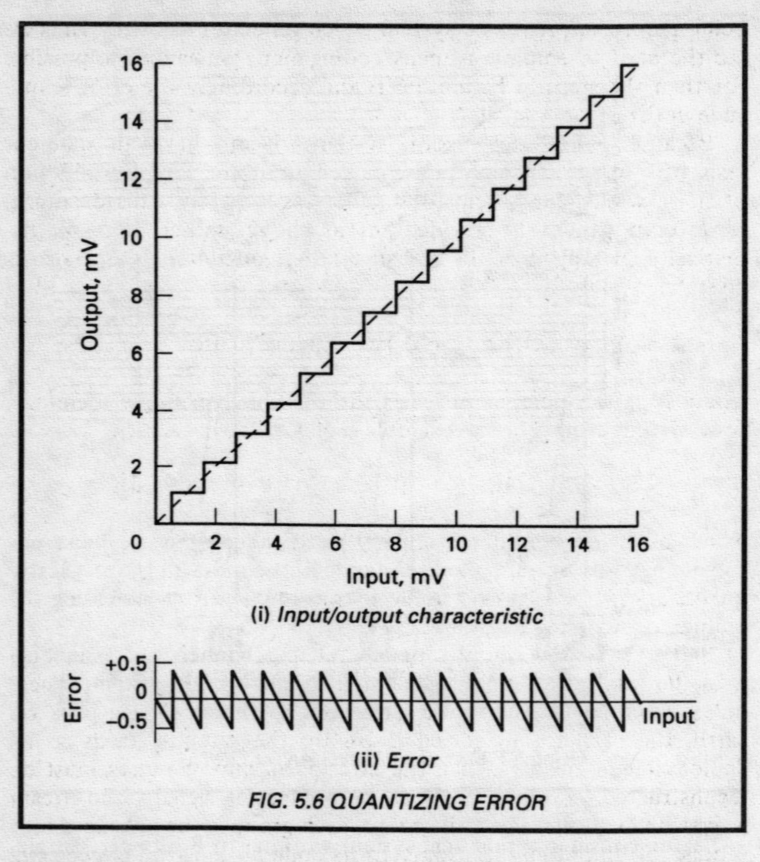

(i) *Input/output characteristic*

(ii) *Error*

FIG. 5.6 QUANTIZING ERROR

instead of 16, hence the accuracy is 16 times better. The Channel 1 sample taken at 125 μs has 4 as the nearest level and again at 250 μs it is 7. Channel 2 in the figure is sampled at 25, 150 and 275 μs resulting in sampling levels of 3, 11, and 10 respectively.

Obviously there must be some impairment from the use of discrete sampling levels otherwise even fewer could be used. Consider Figure 5.6(i) which shows the input/output characteristic of a quantized system as in Figure 5.5. The ideal characteristic is a straight line but the practical characteristic has a staircase form. As an example, for all inputs between just over 0.5 and 1.5, the output is consistently 1. In (ii) of the figure we plot the actual error, this is the *quantizing error* and it gives rise to a system noise known as *quantization noise*. This noise is generated by the p.c.m. system itself. There is always the problem of unwanted noise being picked up, here is a system which generates its own! Thus we see the need to employ as many coding digits per sample as possible for then the steps in Figure 5.6(i) and accordingly the error amplitude in (ii) become smaller.

Because pulse transmission systems overall have an inherent resistance to the effects of noise picked up on the way, introduction of noise which the system itself generates is clearly a disadvantage, however it cannot be avoided for as shown rather more mathematically in Appendix 11 the *signal-to-quantization noise ratio* is given by:

$$\text{SQNR} = (10.79 + 20 \log N)\ \text{dB}$$

where N is the number of levels. In the above example therefore, for a system using 8 bits per sample (i.e. $N = 256$):

$$\text{SQNR} = (10.79 + 20 \log 256)\ \text{dB} = 58.95\ \text{dB}$$

which might be considered as very good indeed but we must not forget that for several systems in tandem, the noise adds up. Clearly quantization noise can be reduced to a low level by increasing the number of sampling levels.

Figure 5.5 shows how separate p.c.m. channels are assembled with the transmitted pulses generated by our simple system. These pulses modulate a carrier wave directed to the satellite. Back on earth, the carrier transmitted down by the satellite needs to be demodulated, meaning that the stream of binary pulses must be reconstructed to form a copy of the originating signal. The stream is first divided into the digit groups, each group is then changed into a pulse of amplitude according to its code by a *digital-to-analogue* converter. Demodulation of a stream of these pulses is then surprisingly easily accomplished by passing them through a low-pass filter which separates the baseband component from the harmonics.

5.6 Frequency Changing

Frequency changing goes on in many aspects of radio transmission, generally moving from a higher frequency to a lower one which can be handled more easily. Implicit in this is the fact that we can change from one carrier frequency to another without affecting the modulating frequencies in any way. A *frequency changer* does just that and there are many instances in satellite transmission where frequency changing is employed. As an example Section 5.3 gives reasons for the use of a higher uplink frequency than for the downlink so indicating that a change of carrier frequency takes place in the satellite itself. Figure 5.7(i) shows how problems arise if no frequency changing is involved. The Figure shows that without a frequency change and highly directional though the antennas may be, some leakage from the powerful downlink signal will be picked up by the receiving antenna. This passes through the amplifier and emerges at an enhanced level so increasing the level of the leakage. The effect is cumulative and the whole system becomes unstable.

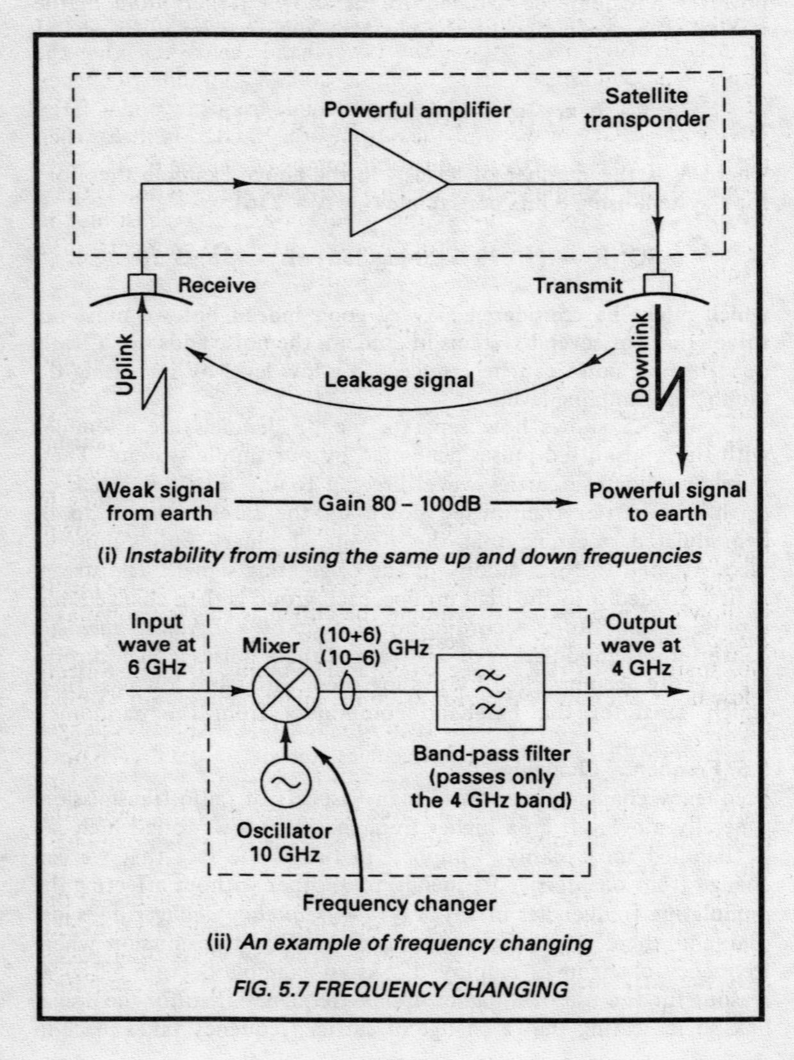

(i) *Instability from using the same up and down frequencies*

(ii) *An example of frequency changing*

FIG. 5.7 FREQUENCY CHANGING

A frequency change within the satellite avoids this because the receiving system does not accept the transmitted frequency.

A frequency changer relies on the fact that if a lower frequency f_1 is mixed in a certain way with a higher frequency f_2, then two other frequencies are produced, $(f_2 + f_1)$ and $(f_2 - f_1)$, the sum and difference. Other frequencies are also there but we get rid of them by filtering. One of the two new components is rejected by use of a *band-pass filter.* This is an electronic circuit within the frequency changer which passes a certain band of frequencies but no others. Suppose it passes the lower band, hence rejecting the upper band and consider that within a satellite an uplink frequency of 6 GHz has to be changed to a downlink frequency of 4 GHz. Mixing the 6 GHz wave with the output of a 10 GHz oscillator then gives (10 – 6) = 4 GHz. Modulation impressed on the 6 GHz wave is carried through onto the 4 GHz so the same information is there but on a carrier wave of a different frequency. This is illustrated in Figure 5.7(ii) which also illustrates the basic components of a frequency changer.

5.7 Transmission Through the Satellite

There are satellites up there solely for gathering information and transmitting it to earth, e.g. for weather and observation. Most communication satellites however operate as relays meaning that they receive information from the ground station and then transmit the same information back down to earth. Frequencies employed are generally in the gigahertz range for reasons given earlier in Section 1.4.2. There is a change of frequency from uplink to downlink within the satellite as shown in the preceding section.

The basic system including the frequency changing arrangements is shown in Figure 5.8. Excluding the antennas this is known as a *transponder* (transmit and respond – er). Several (e.g. 16 – 24 plus spares) are carried by one satellite. The uplink signal first meets a low-noise amplifier which for example, might be based on a tunnel diode. The amplified signal is then directed to a frequency changer which shifts the modulating frequencies (television, data, etc.) onto a lower frequency carrier as shown in the example in Section 5.6. When the output wave is at a lower frequency than the input, the complete unit of mixer, its oscillator and the following band-pass filter is known as a *downconverter.*

As shown in the preceding section, there are two inputs to the mixer, the output signal of which therefore contains the sum and difference frequencies of these inputs. The band-pass filter passes only the difference frequencies. Following the filter a preamplifier

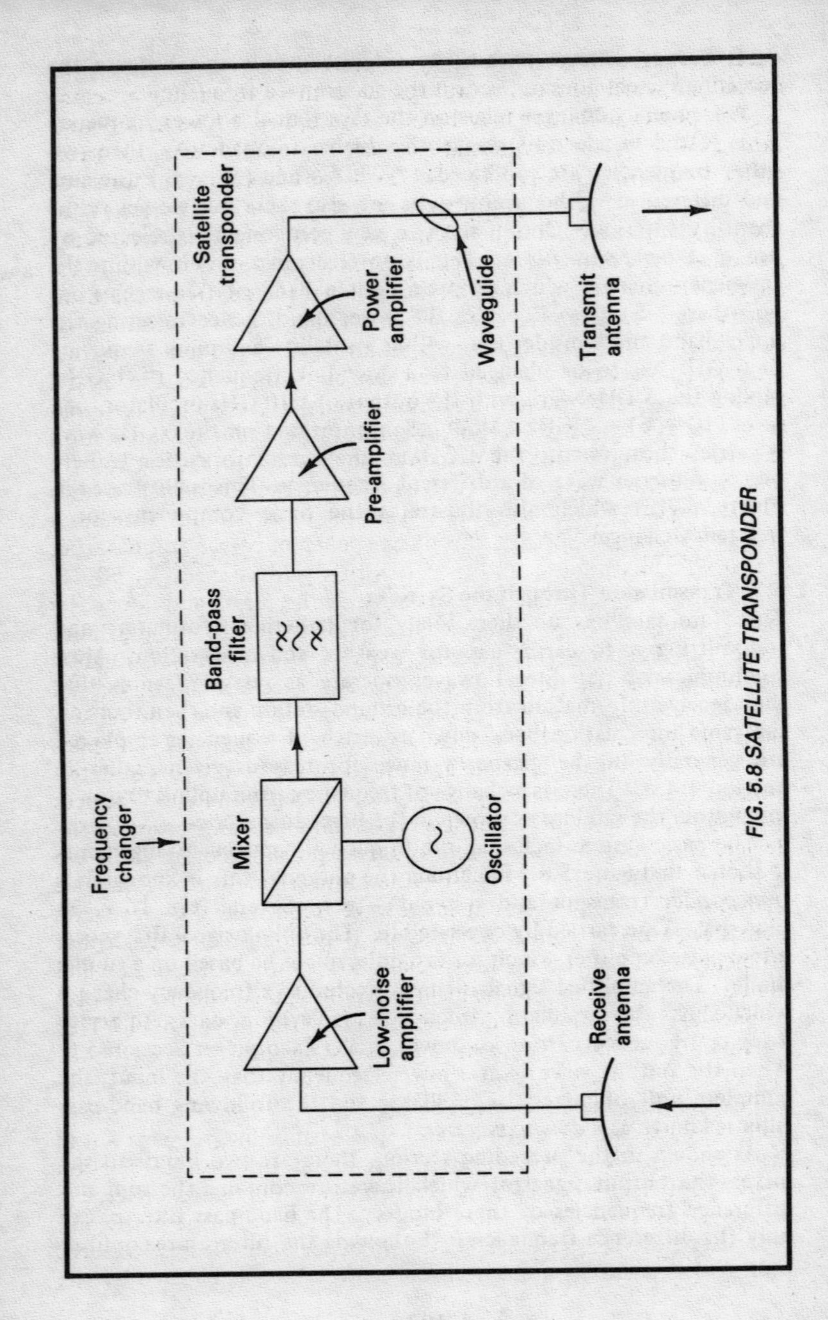

FIG. 5.8 SATELLITE TRANSPONDER

raises the signal level sufficiently to drive the power amplifier, the output of which travels by waveguide to the transmitting antenna.

The power amplifier may be based on either the *travelling-wave tube* (TWT) or on a solid-state device, the *gallium arsenide field-effect transistor* (GaAs FET). The TWT has been in favour for some time because it does the job well, moreover it still is preferred for the higher gigahertz frequencies. In it the electromagnetic wave passes along a kind of waveguide helix within an evacuated glass envelope some 60 cm long. Shooting down the centre of the helix in the same direction as the wave is moving is a narrow stream of electrons. The whole system is so arranged that there is mutual interaction between the wave and the electron stream and as the former progresses around the turns of the helix, energy is transferred from the electron stream to it. By this rather complex interaction the wave is amplified. This could be likened to sliding down a helter-skelter where gravity (the electron stream) continually adds to the motion so that a relatively slow start at the top becomes a speedy exit at the bottom. TWT's can generate power outputs from a few watts to around 200 to 250 with a gain in excess of 50 dB.

Alternatively, gaining in popularity is the GaAs amplifier. Its most useful gains are obtained at the lower end of the gigahertz range although such amplifiers are capable of operating at frequencies as high as 100 GHz. The device is a special FET, very small hence electron transit time is short. In addition the use of gallium arsenide instead of silicon results in an electron mobility 3 – 4 times greater. Electron transit time and mobility are key factors in determining the high frequency response of semiconductor devices because when a high frequency wave initiates action within the device, this must be completed as far as possible before the wave polarity reverses. The GaAs FET therefore has better performance than normal silicon devices at microwave frequencies. Power GaAs transistors can be connected in Class B push-pull for even higher efficiencies. Their low cost, especially low power consumption and the fact that they can now be included in integrated circuits all point to the fact that they are destined to take over from the TWT as satellite power amplifiers in the not too distant future.

5.8 Scrambling and Encryption

Here we have two words often accepted as meaning the same but in fact technically they are different. A better title for this Section would have been *cryptology* (*crypto* meaning concealed or secret). However this is perhaps a term less well understood generally and especially now that satellite television has become well established. Codes, that is signals used to ensure secrecy have been used by the military since time immemorial but the fact that a code is required nearly always indicates that an unauthorized somebody somewhere is anxious to read the information being transmitted and therefore is likely to be busy in endeavouring to "crack" the code.

Things have not changed today, codes even more difficult to crack are used by military, national security intelligence services, banks and others requiring the highest level of security. The systems have become more and more complicated simply because those "on the other side", wishing to find out what is going on, also use more sophisticated techniques. Not so many years ago, secret codes were a matter only for the experts but nowadays with the advent of satellite television, we are all becoming involved because broadcasters who make a charge when people watch certain programmes make sure that only those who have paid up are able to watch and so a decoder has to be fitted.

The difference between scrambling and encryption, the modern expression for "secret code", may be considered trivial but we should have some appreciation of how they differ. Scrambling can be considered as the method by which frequencies are jumbled in an effort to make the outcome unintelligible except to the intended recipient. Encryption goes one stage further and is a more complex process which requires additional signals or *keys* for *decryption.* Thus the information transmitted includes not only the encrypted signal but also the control data for decryption.

Needless to mention perhaps, is the fact that we are unable to discuss the more complex and secret systems and not at all those for the military and secret services. Nevertheless it is possible to appreciate the basic ideas on which some are based. We therefore look firstly at a digital system in a very simple way, this is one based on digital addition equivalent to an exclusive – OR logic operation. This we recall is an OR gate which excludes the condition of both inputs being at logic 1 simultaneously:

$$0 + 0 = 0 \qquad 0 + 1 = 1 \qquad 1 + 0 = 1 \qquad 1 + 1 = 0$$

Suppose we require to transmit the character Z as an 8-bit binary code 01011010 and the encryption key to be used is 10101010. The process is then as follows:

character Z in binary	01011010	
encryption key	10101010	now add
transmit	11110000	

Assume this is received correctly then add

encryption key	10101010
to obtain the transmitted code	01011010 by binary addition.

This is the original character (Z) but not recognizable as such on the channel. This system as demonstrated while providing some security fails when the key is discovered by others. However it can be made considerably more secure by increasing the number of encryption keys used in sequence. Remember this is only a single, simple example, many are in use of greater complexity according to the degree of security required.

One of the present systems for the encryption of television signals is aptly known as *Videocrypt*, it is one of the systems finding favour and is now widely used in the UK. We recall that the requirement of the system is to limit viewing to those subscribers who are paying for the service. One of the problems the operator has to endure arises from the fact that "pirate" decoders are not illegal and much literature is available on "hacking". Hackers are people with sufficient electronics skills to be able to design their own decoders either for financial gain (e.g. breaking into banking systems) or alternatively for the sheer delight of success.

The basic principles of television are discussed later in Chapter 7. All we need to know here is that the picture is built up by a series of lines as is this page. Taking as an example, a 625 line system, only 585 of these lines are used to produce the visible picture, the time equivalent to the remaining 40 lines being used to transmit control (e.g. picture synchronization) and teletext data.

In the Videocrypt system each horizontal line of the picture is cut somewhere and the parts of the line on both sides of the cut are changed over, i.e. the right-hand side of the cut line becomes the left (the so-called "cut and rotate" system). Such a system would be a joy to a hacker except for the fact that the position of the cut in each line is different. Clearly then, at the receiving end the decoder must be advised as to where the cut has been made in each line. This *key* information is sent to the receiver within the 40 line interval mentioned above similar to the way in which teletext is transmitted as a digital code. But this is not all, the final part of the key is stored in the magnetic memory on the viewer's *smart card.* This is similar to a modern credit card with its magnetic stripe. Difficult to appreciate perhaps but the smart card actually embodies a tiny computer and memory chip. The system enables all decoders to be the same but each smart card matches not only the transmitted signals but also the viewer. The transmitted codes are changed regularly and at each change viewers are issued with new cards. The hacker is therefore confronted with a system which is difficult enough to crack, say by reading the code on an authorized card, but the work is rendered useless immediately the transmitted key code is changed. The Videocrypt equipment may be built into the satellite receiver or alternatively assembled as a separate unit.

One might be forgiven for doubting whether with all the manipulation of the picture signal which goes on, the picture quality would be degraded. It is a tribute to modern technical achievement that although there is in fact some slight degradation, this is so small as to be unnoticeable.

Chapter 6

TELEPHONY AND DATA SYSTEMS

Undoubtedly in the early days of satellite experiments their use for the transmission of telephony and data was eagerly awaited. Satellites had the potential to span earthly distances of 1000 – 2000 km (or even more) as easily as over a distance of just a few kilometres. Moreover the oceans would be no barrier. Nowadays all this has been achieved so that whereas an extensive network of underground cables, microwave links and undersea cables may be required for long-distance or international working, the satellite can achieve the same in one "hop". This is illustrated in Figure 6.1 which shows a particular connection between two telephone subscribers. It is clear that an extensive land-based network of cables, microwave links, etc. is required to provide the same wide area coverage as that of a single satellite.

Lest we become too enthusiastic about the advantages of satellite telephony, we are reminded that modern fibre-optic cables across both land and sea equally have much to offer, even cost-wise. There is a further imperfection in the satellite system which no amount of research or development can cure, that of transmission time.

6.1 Transmission Time

Although the electromagnetic wave travels so fast that it can encircle the earth seven times in only one second, the fact that satellites are far above the equator (say, 36 000 km) means that one round trip from *A* to *B* (Fig.6.1) will take:

$$\frac{2 \times 36\,000 \times 1000}{3 \times 10^8} = 0.24 \text{ seconds} = 240 \text{ milliseconds.}$$

This is the shortest time, i.e. for both stations directly below the satellite. On average the figure becomes about 270 ms, twice this between speaking and receipt of an answer, that is, over half a second. This is just tolerable but the delay for two satellite links in tandem is not, it causes confusion in conversation. Hence international calls do not contain more than one satellite hop and unfortunately as indicated above, nothing can be done to shorten the time for we cannot make electromagnetic waves go any faster! Thus although for the single hop, telephone calls by satellite are

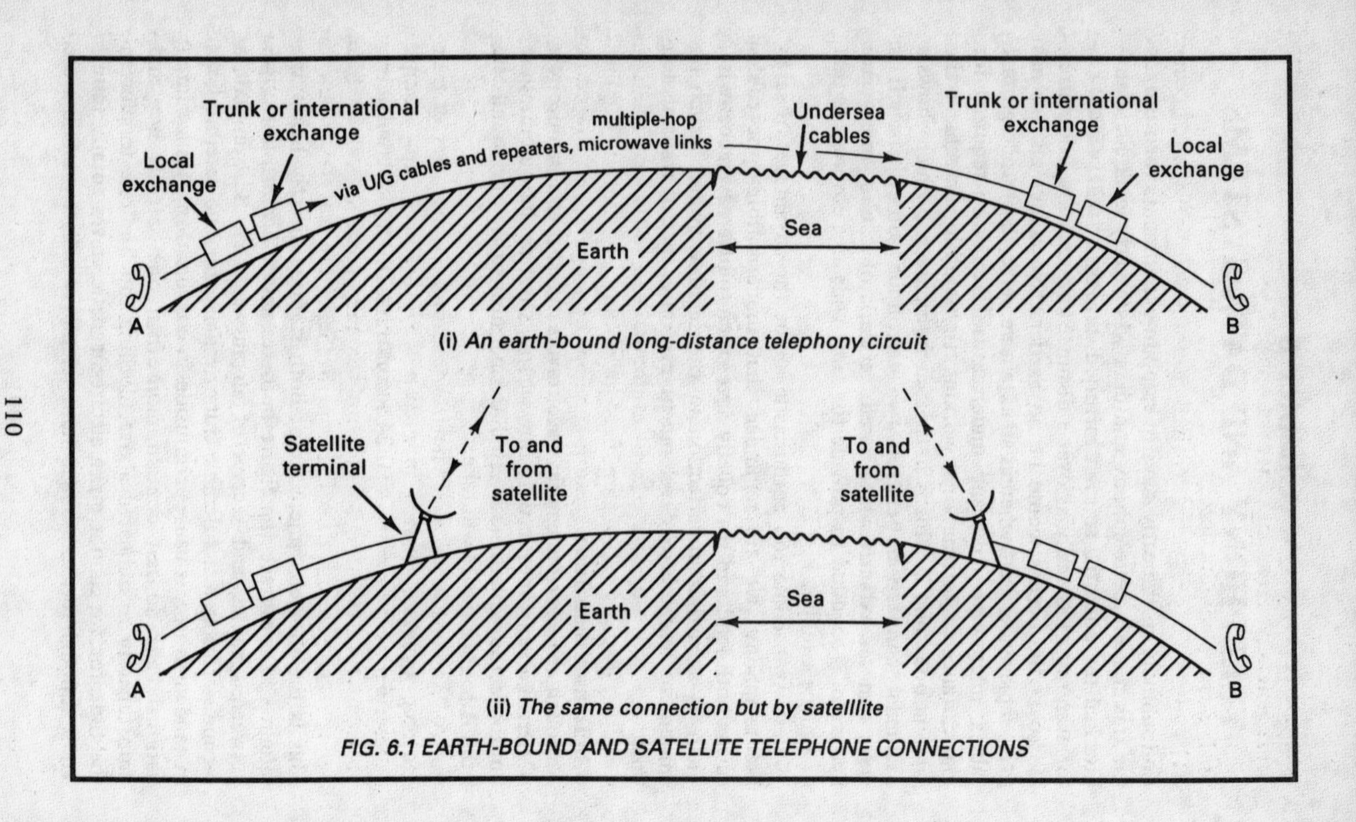

(i) An earth-bound long-distance telephony circuit

(ii) The same connection but by satelllite

FIG. 6.1 EARTH-BOUND AND SATELLITE TELEPHONE CONNECTIONS

considered acceptable by some 90% of users, it must be admitted that generally terrestrial systems are slightly better.

6.2 Signal Reproduction

The quality of the signal received over any telephony link depends on many factors but generally telephony circuits, although usually of restricted frequency range (300 – 3400 Hz) leave little to be desired as far as quality is concerned, bearing in mind that we are considering "commercial" speech, not "high quality". Noise and echo, both of which detract from good speech quality, are probably the most troublesome conditions which arise, especially on long-distance circuits. Noise is considered in Chapter 5 but echo needs special attention for it becomes appreciable on long circuits and as we have seen in Section 6.1 satellite circuits are certainly very long. Echo is present on any telephony circuit and it becomes objectionable when a talker hears an echo of the speech delayed by more than just a few milliseconds.

6.2.1 Echo

Echo arises mainly from the return of a signal from an imperfectly balanced terminating set at the distant end, delayed by the double journey transmission time as illustrated in Figure 6.2. Terminating sets (also known as *hybrid transformers*) are required in any long-distance and therefore amplified circuit because the amplifiers are one-way devices. The amplifiers have input and output terminals and a signal applied at the input appears amplified at the output but amplification does not take place in the reverse direction. There are exceptions to this, e.g. with negative impedance amplifiers but these are not used generally because they have their own problems.

A terminating set is a Wheatstone Bridge arrangement of transformer windings coupled in such a way that when for example, subscriber *A* talks or data is being transmitted, voice-frequency currents flow into the 2-wire terminals and appear at both the Send and Receive terminals. That on the Send line travels by land or satellite link to the terminating set at *B*. Here the signal arriving at the terminating set Receive terminals crosses over to the 2-wire terminals for reception by the telephone at *B*. Transmission of speech or data from *B* to *A* is by similar means. Unfortunately it is not as simple as it appears. Unless a terminating set is accurately balanced impedance-wise to the 2-wire line and its termination, there is a leakage of signal from Receive to Send. The loop gain of the whole circuit is in effect the difference between the total amplification and the total losses (including those of the terminating sets).

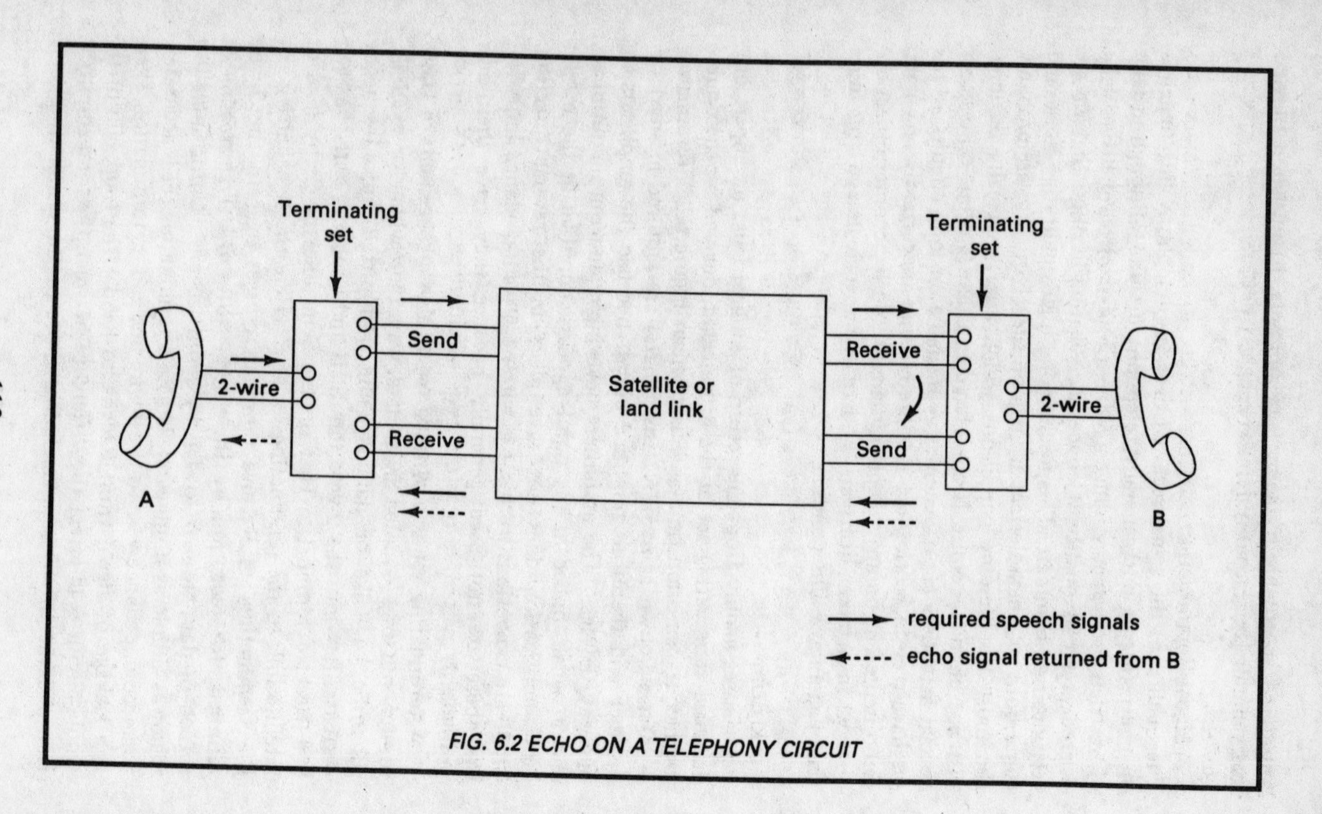

FIG. 6.2 ECHO ON A TELEPHONY CIRCUIT

If the loop gain exceeds zero at any particular frequency then the whole circuit becomes unstable because in fact the system is now an oscillator.

Accordingly some of the signal arriving at the terminating set at *B* returns to *A*, this is the echo which is heard by *A*, delayed by the total time the signal has travelled to the distant end (*B*) and back again. Clearly when a satellite connection is involved, the echo will occur delayed by the double journey transmission time, i.e. at least half a second later. Early experiments soon indicated that echoes of this nature even when received at a considerably lower level compared with the originating speech were disconcerting to the speaker and were therefore not tolerable. As one might imagine, the degree of disturbance by the echo signal is related both to its delay and to its magnitude. Accordingly *echo suppressors* are employed on all such circuits.

Echo suppressors are installed in the 4-wire section of a circuit. They operate by introducing a loss in one path on detecting a signal in the other. It is clear that they must operate and release quickly otherwise speech in either direction will be lost. Here we are of course talking in terms of a few milliseconds. In Figure 6.2 therefore any utterance by *A* is detected on the Send line, so resulting in the insertion of attenuation in the Receive line. As soon as *A* is quiet the circuit reopens so that speech from *B* can be received. This all sounds very simple but in fact human conversation is anything but disciplined hence arrangements are built in so that either party is able to override the other. Altogether a switching arrangement for the suppression of echo, made complex by the need to ensure that there is little or no embarrassment to normal telephone conversation.

Digital transmission over this type of circuit now enjoys superior echo control which has been more recently developed and is known as an *echo canceller.* The system employs a microprocessor operating with high speed digital circuits, needless to say, the echo suppression achieved exceeds that of the more common suppressor described above.

6.3 Companding

This is a technique used for the improvement of signal-to-noise ratio over audio frequency or multiplex circuits. Basically it reforms the dynamic range of a sound channel by:

(i) at the transmitting end more gain is given to low amplitude frequencies than to high amplitude ones (the *volume compressor*);

(ii) at the receiving end the signal is expanded in the opposite sense to restore it to its original range of amplitudes (the *volume expander*).

Low level input signals are therefore less affected by noise picked up on the channel because of their increased amplitudes, i.e. there is an improved signal-to-noise ratio. When these frequencies are reduced to their original amplitudes in the expander, the noise is similarly reduced. The signal level over the link therefore is deliberately kept fairly high so maintaining it well above the noise.

Chapter 7
TELEVISION BROADCASTING

In the foregoing chapters many of the basic concepts on which satellite television is founded are examined. In this chapter we attempt to build on them in order to obtain a better acquaintance with the whole system with a view perhaps to installing or at least maintaining our own home installation. Figure 7.1(i) shows the basic system, the transmitting station sends the television signals over a narrow radio beam to the satellite at a frequency between say, 11 and 14 GHz (4 – 6 GHz systems are used mainly in the USA). The signals are received on a satellite parabolic antenna pointing directly down the uplink beam. Within the satellite the transmission frequency (not the modulation – see Sect.5.7) is reduced for transmission over the downlink to the area served on earth. The downlink is shared by the various types of users as shown in Figure 7.1(ii). The classifications of the receiving terminals shown in the figure are:

CATV = Community Antenna Television
SMATV = Satellite to Master Antenna Television
TVRO = Television Receive Only

(see also Appendix 1). What we hope to achieve in this Chapter is a better understanding of the technicalities involved in the home environment and how we can get the best out of our own home installation.

A drawing of a complete home system not only for terrestrial (earth-bound) reception but also for satellite is given in Figure 7.2. At present because terrestrial television has long been with us, we must look upon satellite television as a late arrival providing an added facility. There may come a time when everything for both terrestrial and satellite television is built into one receiver and it has even been mooted that satellite television will ultimately replace terrestrial altogether. At present however most installations consist of a standard television set with the add-on equipment for satellite reception as shown.

If some revision about antennas is required we must return to Chapter 4 while the decoder is considered in Section 5.8 and note that as time progresses, because broadcasters have now made up their minds as to the decoding system each requires, then probably the decoder will be built into the satellite receiver. The terrestrial

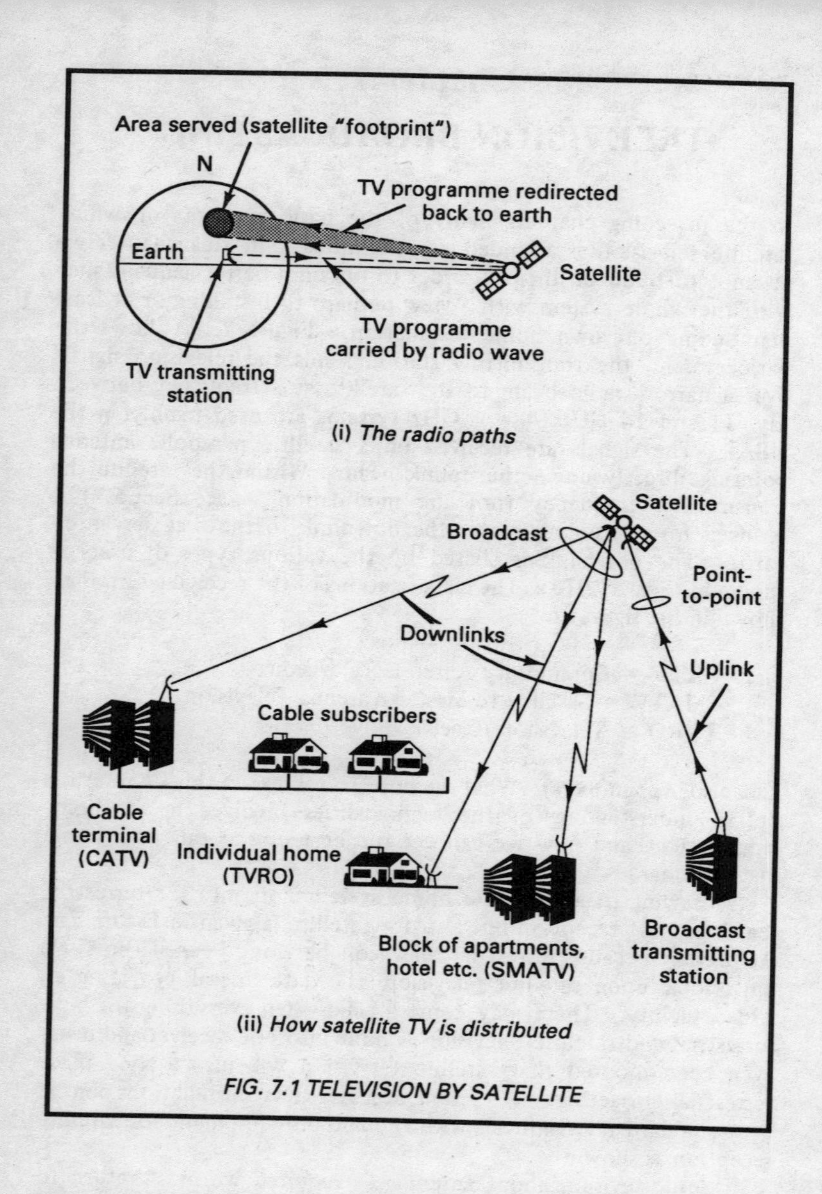

FIG. 7.1 TELEVISION BY SATELLITE

antenna is shown connected directly to the television set. In practice it is more likely to be connected to and switched by the satellite receiver or the video recorder. However firstly we ought to get to grips with the television receiver itself.

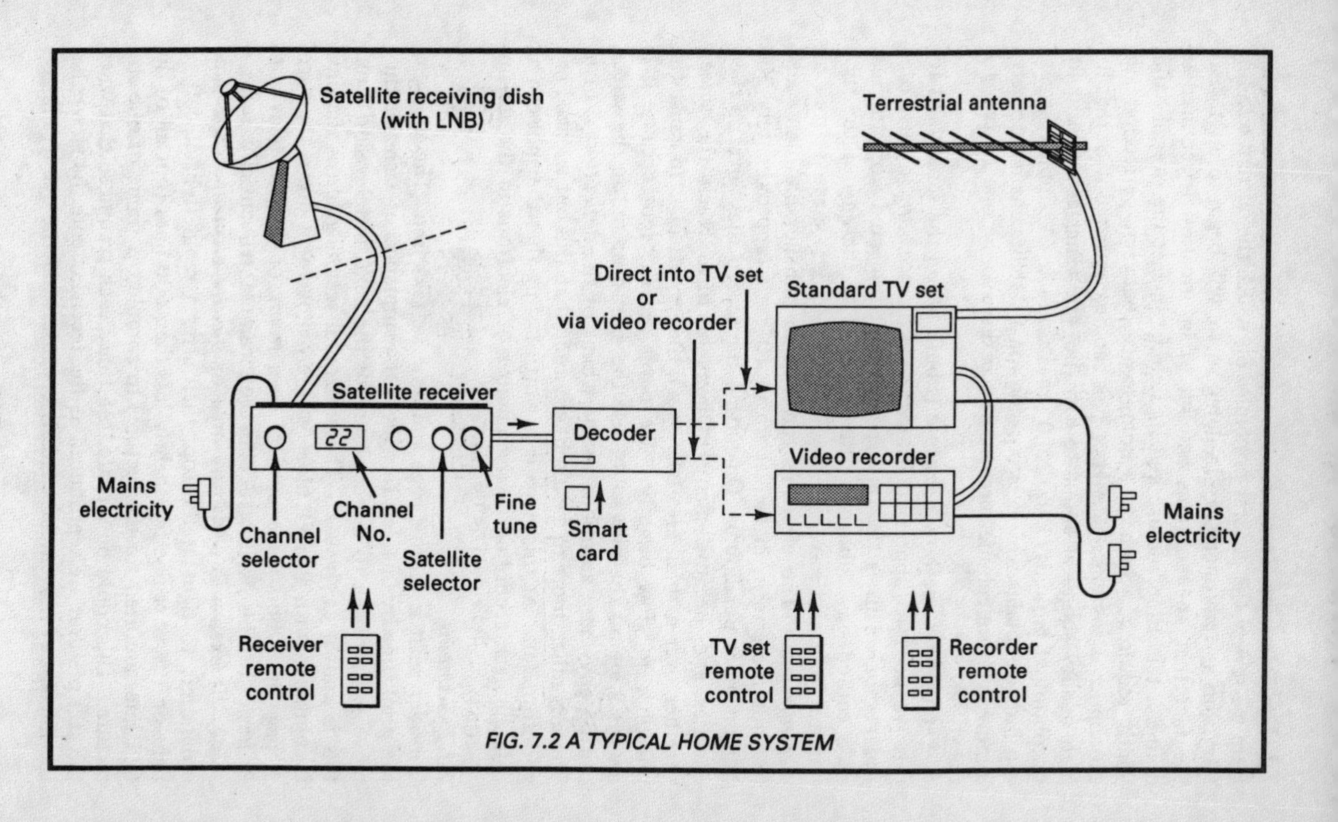

FIG. 7.2 A TYPICAL HOME SYSTEM

7.1 The Television Receiver

Here we review the general principles on which the current 625-line television system operates. In doing so we will begin to appreciate the fact that an enormous amount of information is required to define a television picture and its sound. In fact the overall bandwidth required, which is a compromise between perfection and economics is some 6 – 7 MHz. This can be compared with that for high quality sound at about 15 kHz, i.e. 400 times as much or for telephony at around 3 kHz, 2000 times as much. A simple comparison but it explains why it is possible for a satellite channel to carry hundreds of telephone circuits but perhaps only one or two for television.

Let us next see how a picture is built up on the screen and for this we need to appreciate the method of *scanning* by use of a *flying spot.* This is the tiny spot of light projected from the rear of a cathode-ray tube onto the screen. At the end of the tube remote from the screen a beam of electrons is ejected from an *electron gun.* The electrons travel at an unbelievable high speed towards the screen, in fact at up to 10^8 m/s (approximately 200 million miles per hour). On hitting the screen they give up their energies and because many millions of them reach the screen together over a small area (the spot), the net effect is to cause the phosphor to glow brightly. We never see the spot because it is always kept on the move. In use the spot position on the screen can be anywhere according to the whims of the *deflection system.* There is much more to a cathode-ray tube than this but we need not go into detail. At this stage it is worth noting that the spot can be light or dark according to the number of electrons in the stream.

7.1.1 Scanning

Consider what is on this page and how we take it in. The eyes scan the print, line by line, left to right. On completion of each line they move swiftly back to the beginning of the next, in television terms this is known as *flyback.* The horizontal scanning across the lines is combined with a much more leisurely movement down the page. A page in a book is equivalent to a *frame* or *field* in television.

Basically this is how a television camera sequentially scans a scene. It examines each tiny area or *picture element* in the same order, left to right at the top, then after a rapid flyback, left to right on the next line for slightly less than 600 lines (not all of the 625 lines are used for the picture) to complete a frame of one still picture. This could be repeated 25 times each second so that when the still pictures are reproduced at the receiving end, the viewer has the illusion of movement. This is because what the eye sees lingers on for a short time (persistence of vision) so each full picture appears to merge with the previous one. Unfortunately 25 pictures per second results in a noticeable *flicker* for most viewers. On the other hand 50 pictures per second does not.

Normally doubling the number of pictures per second would double the information rate and therefore the bandwidth, an expensive commodity in satellite transmission. A technique of *interlaced scanning* is used instead. Rather than transmitting each line in sequence, the odd lines are scanned for one frame and the even for the next, giving two half-scans per picture. The effective presentation therefore is at 50 images per second with flicker reduced accordingly.

Considering next the home television screen, if the tiny spot of light is controlled in this way, the result is a bright white rectangle the size of the screen. The pattern of scanning lines is known as a *raster.* An electronic *line time base* moves the spot across the screen and then lets it fly back so quickly that it is not seen. At the same time a *frame time base* is moving the spot downwards. After a frame is completed the spot returns to the top and while this is happening, because no picture information is being received, the transmitter seizes the opportunity to send *teletext* and other data.

7.1.2 Synchronization

When a television camera is scanning a scene and television receivers are reproducing it, their scanning arrangements must work in unison, i.e. when the camera scanner commences a frame, the receivers must also. Cameras and receivers must therefore be in *synchronism.* The system does even more, not only do the frames start in synchronism but so does every line. This is achieved by sending a short duration *triggering pulse* which is a tiny burst of energy causing the line time base to start drawing the spot across the screen. The pulse is only about 5 μs long, occurring every 64 μs (the time taken for one complete line). Hence when the camera commences scanning a particular line, all television sets receiving the programme start to reproduce the same line. When the line is completed, all spots fly back to the left to await the next triggering pulse for starting the line below. Similar arrangements shift all spots from the bottom of the screen to the top to await triggering for the start of the next frame.

7.1.3 The Picture

A device which adjusts the number of electrons in the beam which terminates as a spot of light on the screen is all that is required for a monochrome picture. We call this adjusting the brightness or *luminance.* With no electrons, there is no spot, i.e. black, with the full quota there is white. In between are all the shades of grey. Provision of this facility is quite simple. Within the cathode-ray tube a *grid* in the *gun* which generates the electron beam from the rear, surrounds the electron beam just after it is generated. When the camera sees white the signal it sends out arrives on the grid and allows the full electron beam to pass. Anything less than white puts a negative potential on the grid so repelling electron flow and producing a less than full-white spot. Summing up therefore and considering a single line of the picture, it is the synchronizing pulse which arrives first and this is directed to the line time base which it starts. The remaining incoming waveform passes to the grid of the cathode-ray tube to control the brightness of the spot as it moves across the screen.

This is for a black and white picture. Things really get complicated when colour is added. Fortunately any colour can be made up by mixing together any of three primary colours only – red, green and blue (R, G, B). In essence the camera separates light from the scene into these primaries and directs each to a special pick-up tube. The electrical outputs from the tubes are mixed in such a way that they can be separated later and the mixture forms the *chrominance* (colour) waveform. This is transmitted together with the luminance, synchronizing pulses and audio signals. The elements of the whole system are shown in Figure 7.3 which is discussed in more detail below.

In the receiving cathode-ray tube there are now three electron guns, one for each primary colour. The screen fluorescent coating is more complicated than for monochrome. It is coated with microscopic phosphor dots in groups of three. When any dot of a group is hit by an electron beam it glows in its own colour. A special perforated mask ensures that electrons from each gun strike the appropriate phosphor dot, e.g. those from the electron gun for red are directed only to the phosphor dots which glow red. Each phosphor dot in a group therefore contributes an amount of its colour as determined by the strength of its electron beam. As an example, illuminating the red and green phosphor dots but not the blue in a group produces yellow. Adding various small amounts of blue gives a range of pastel greens (faintly exciting the green dot on its own gives a pastel green but of the basic shade only).

Returning now to Figure 7.3, signals sent down by the satellite are received and selected by the satellite receiver (see Sect.7.2.2). Generally the satellite receiver shifts the incoming carrier frequency to that of the channel to which the television set is tuned specifically for the purpose of receiving satellite transmissions. In the UK this is

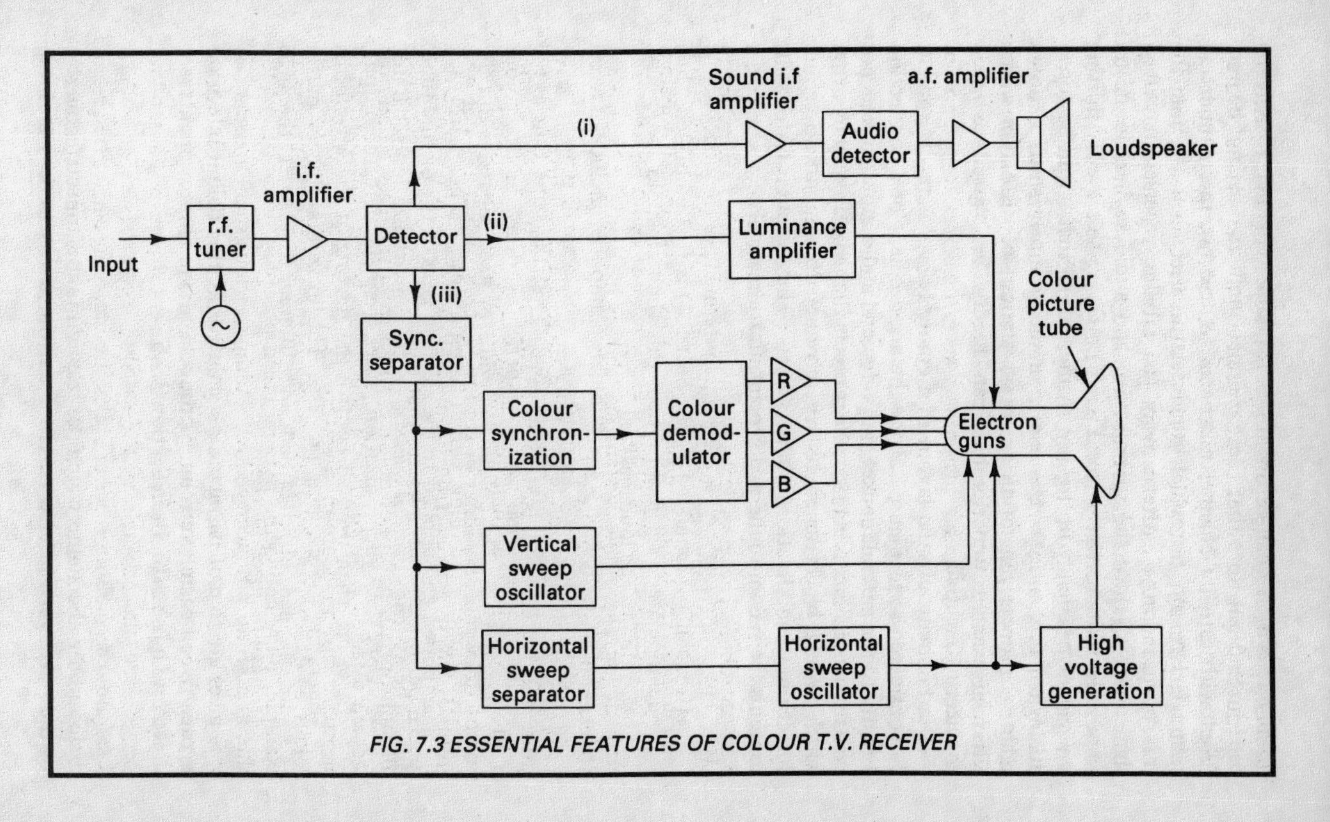

FIG. 7.3 ESSENTIAL FEATURES OF COLOUR T.V. RECEIVER

usually television channel 36 at 591.25 MHz (channel 38 may be used instead but note that these channel numbers must not be confused with satellite channel numbers). These two interconnection channels have no terrestrial broadcasts on them and are used for this purpose only. Alternatively the satellite receiver may feed signals directly into the special video and audio sockets found on some television receivers (see Fig.7.6). Doing this avoids shifting the carrier frequency up to, say channel 36 so that it can be tuned in. When this is done, the television set promptly shifts it down again, a process generating a certain amount of distortion which although usually unnoticeable, can hardly be classed as good engineering practice.

Diagrammatically the layout of a colour television receiver might be considered as in Figure 7.3 for the arrangement of a connection directly to the antenna socket (i.e. video and audio input sockets not available). The incoming signal first meets the radio frequency tuner which would be switched to say, channel 36, amplification and detection (demodulation) then follow. The output of this unit contains the vision, sound and synchronization signals.

(i) The *sound channel* – the sound intermediate frequency signal (frequency modulated – see Sect.5.4) is first amplified, then demodulated, again amplified, this time at audio frequency, then fed to the loudspeaker.

(ii) The *video section* – the luminance amplifier includes several stages of wide-band amplification with a small time delay so that the information reaches the tube at the same time as the colour information which experiences delay in the colour demodulator. The colour demodulator itself separates out the red, green and blue colour components. These are each amplified and applied to the electron guns of the picture tube.

(iii) The *synchronization section* – the synchronization separator accepts the synchronization pulses from the detector output, these are then used to control the starting of the waveforms operating the scanning process. The horizontal flyback pulses are stepped up, rectified and filtered to provide the high voltage required by the tube.

It is a tribute to modern electronic engineering that colour television is capable of such exquisite reproduction, yet improvements are already with us as shown in Section 7.1.5.

7.1.4 Sound

Undoubtedly the sound output of modern television sets is moving towards the stereophonic (*stereo* from the Greek *stereos* meaning solid or in our way of thinking, three dimensional). Satellite broadcasts cater for this so most satellite receivers are designed accordingly. In the studio, for stereo two or more spaced microphones are used working into completely separate sound channels which ultimately feed individual loudspeakers spaced apart in the home. By this means some of the spatial effect in the original sound is transmitted to the listener. Generally only two sound channels are employed but the overall effect of greater realism can be impressive. Nevertheless *monophonic* sound is always there as a high deviation frequency modulated signal (Sect.5.4) employing a bandwidth of almost 300 kHz. On ASTRA the mono sound is on a 6.5 MHz subcarrier.

Most of us when discussing high fidelity (hi-fi) reproduction are inclined to think mainly in terms of frequency response and freedom from distortion as is available from a cassette disc (CD). When a radio link is included in the chain however, noise must also be considered for there is nothing hi-fi about a signal if it is swamped by noise which has been added during the journey. That brings us back to the all-important signal-to-noise ratio.

A successful and well known sound stereo system has been developed by *Wegener*, an American company. It is known under the general name *PANDA* (from *Processed Narrow Deviation Audio*). This system combines standard companding (Sect.6.3) with additional pre-emphasis on transmitting which boosts the higher frequencies since they are more likely to suffer from channel noise than the lower ones. In the receiver de-emphasis restores the signal to its original form. The net effect is that the channel certainly provides hi-fi quality.

NICAM is a digital stereo system. It is an acronym of *Near Instantaneous Companded Audio Multiplex* and the system converts analogue stereophonic signals to digital and then processes them for transmission with the television picture. Looking at its rather complicated title, *near instantaneous* refers to the high speed at which the analogue to digital conversion takes place; *companding* is the process by which the transmitted signal is compressed before transmission and then at the receiving end expanded to its original form; *multiplex* indicates that the transmitted stereo signal is mixed with other data. *Audio*, of course, refers to the fact that the input and output signals are in the (high quality) audio range.

Sampling of both the left and right audio channels is at 32 times per millisecond so at 11 bits per sample, for the two channels there is a total of 704 bits. With 24 bits added for control data, this generates a 728 bit frame. This is transmitted in the separate sound channel on a centre frequency above those of the picture and mono signals. Being a digital system, NICAM not only results in high quality reproduction (provided that the original source material is also of high quality) but also background noise is virtually non-existent.

7.1.5 A New Look With MAC

Whereas most viewers would regard the existing television in the UK as good enough, the more fastidious may not be so satisfied. The present colour television standards have been with us for some 30 years and most homes have receivers designed to them. It is simply a fact that our present system was originally designed for black and white pictures, then later adapted for colour – it was never designed as a colour system. The introduction of change is therefore not a simple matter. However technology has advanced considerably during this period and we now have the chance to introduce improvements as is happening elsewhere in Europe.

The main defects in the older style television transmissions arise from the way the video signal is put together. Perhaps the most noticeable are due to *cross colour* and *cross luminance.* The first shows itself as spurious swirling coloured patterns over parts of the picture containing fine detail such as narrow stripes on clothing. Although most viewers have become accustomed to the effect and ignore it, the picture is not a faithful reproduction and would be improved without it. Cross luminance shows itself as variations in brightness where extreme colour changes occur, resulting in tiny shimmering dots along the line of colour change, especially at vertical boundaries.

These difficulties arise from the fact that the luminance and chrominance signals share the same frequency band, in a way therefore they are mixed together. In the television receiver these components have to be separated but with the PAL system as at present used in the UK this cannot be done completely, hence there is a certain degradation of picture quality. (PAL stands for Phase-Alternating Line, a concept we need not grapple with here.)

Considerable research has been undertaken by more than one organization to improve the picture quality and the system considered to be technically superior and generally accepted for Europe was brought to fruition in the early 1980's. The main change is that the video signal no longer contains a frequency mixture of luminance and chrominance information, instead these are separated in time. Thus for each television line, the two arrive in "packets" sent one after the other. The system is known as "Multiplexed Analogue Components" (MAC), in fact describing just what it does, the word "multiplex" indicating the transmission of several *separate*

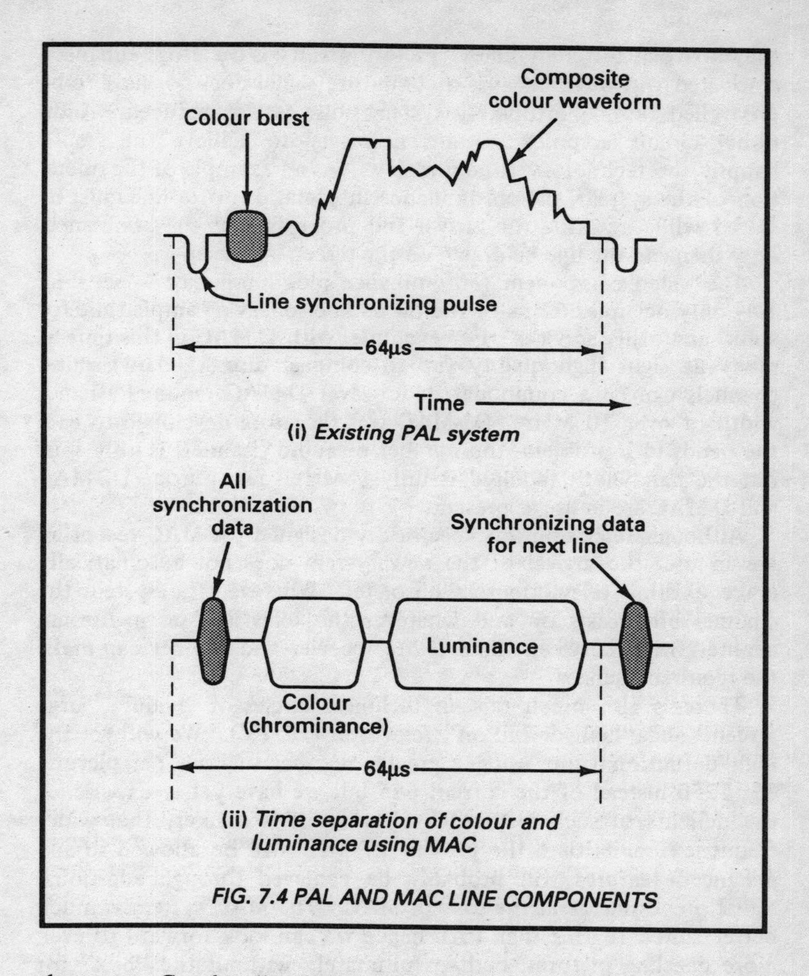

FIG. 7.4 PAL AND MAC LINE COMPONENTS

elements. Graphically the system is best illustrated on a basis of time as in Figure 7.4 which illustrates the fundamental differences between the PAL and MAC systems.

In both (i) and (ii) of the figure we see a *colour burst.* This is used to synchronize the colour oscillator which is used to decode the colour information for each line scan. In the case of the PAL line signal the waveform at any instant after the synchronizing pulse and colour burst represents the total amplitude of the luminance and chrominance frequencies which control the spot as it moves across the screen. In (ii) of the figure it is evident that for the MAC line signal the luminance and chrominance cannot mix because they arrive at different times. Picture quality is therefore enhanced compared with PAL because a composite signal does not have to be unravelled first. Fortunately system noise is also reduced. Additional circuit complexities are necessary to achieve the result, happily the technology is now with us. As an example of the operation of the system, the chrominance information for a line must be stored while awaiting the arrival and processing of the luminance, only then can the line be drawn on the screen by the spot.

The video component (chrominance plus luminance – see Fig. 7.4) only occupies 52 μs of the 64 μs line so leaving ample time for audio and other services. For example, with D-MAC in this time as many as eight high quality (up to compact disc standard) sound channels can be accommodated, however D-MAC requires a bandwidth of over 10 MHz. D2-MAC was therefore developed to ease the bandwidth problem, the number of audio channels is only four but the bandwidth required is only 7 MHz. For Europe D2-MAC and D-MAC are in use at present.

Although television sets specifically designed for MAC reception are in use, the arrival of the new system does not automatically make existing television sets obsolete. Whatever the system the channel broadcasts on and whatever the television set in use, an adapter fitted between the satellite receiver and the set can make the required changes.

There is also much talk in technical circles of "wide", "large screen" and "high definition" television (HDTV). We will get the high definition from using a greater number of lines per picture, e.g. 1250 instead of the current 625 but we have yet to experience the delights of such a system. However it is unlikely that wider frequency bands than the present 27 MHz will be allowed so any enhanced features will probably be achieved through additional signal processing as in the case of MAC. The MAC system is much better suited to this than PAL hence we can look forward to even more dazzling pictures, perhaps ultimately without the "box" for the much delayed flat screen (i.e. with no lengthy tube behind it) must surely arrive one day.

The MAC system uses digital transmission for its sound signals (Sect.5.5) so giving rise to improvements in sound quality. Because digital signals are reconstituted in the receiver "as new", the received signal-to-noise ratio is identical with the transmitted one, noise along the channel having no effect. The actual sound system in use is indicated in the code preceding the letters MAC.

7.2 From Dish to Television Set

In this section we look more closely at the equipment which must be added to a standard television set for the reception of satellite television signals. Chapter 4 considers antennas for reception generally and in Section 4.8 explains the principle of the motorized dish. However before discussing the satellite receiver we must discover what the low-noise block converter does for it seems to be a small yet complex box or tube of equipment always situated on the antenna and for some reason, out in the open where rain and corrosion abound in plenty.

7.2.1 The Low-Noise Block Converter (LNB)

LNB's are shown in Figure 4.2, they are accurately positioned at the focus of a reflecting antenna (this excludes the flat antenna – see Sect.4.2.3). The LNB is probably the most important item of equipment in the receiving installation for without an efficient LNB the whole system is at risk. Placing it at the focus of a home dish avoids the extra expense and inconvenience of a long length of waveguide (Sect.1.4.5). It has two functions: (i) to accept the weak incoming signals reflected from the dish surface; and then (ii), to amplify and convert them to a lower frequency for transmission over a special cable into the home.

In Section 1.4.5 it is shown that radio waves can be enclosed and travel within a tube or guide. The *feedhorn* of the LNB (see Fig. 7.5) is a specially shaped device fitted to a short section of waveguide from which waves can be projected onto the dish for transmitting or equally collected from a dish when receiving. In the drawing a pyramidal feedhorn is shown, it can equally be of conical shape. The dimensions of the horn are controlled by the range of wavelengths with which it is to be used. In Figure 7.5 the body of the LNB contains the electronics – a low-noise amplifier (there is a discussion on noise in Sect.1.5) and the block converter. The latter is a frequency-changer (Sect.5.6) which accepts the incoming band (or block) of signals and then changes them to a similar band but now centralized on a lower frequency. If this were not done, signals at around 12 GHz would need to be fed by a waveguide directly into the home. Use of cable instead at this frequency is not practicable because cable losses increase with frequency and for 12 GHz a cable would be very expensive indeed. It is better therefore to lower the frequency first so that cable can be employed instead of a waveguide. The operating characteristics of LNB's are quoted by the manufacturers and one important feature which should not be missed is the frequency range over which the LNB operates.

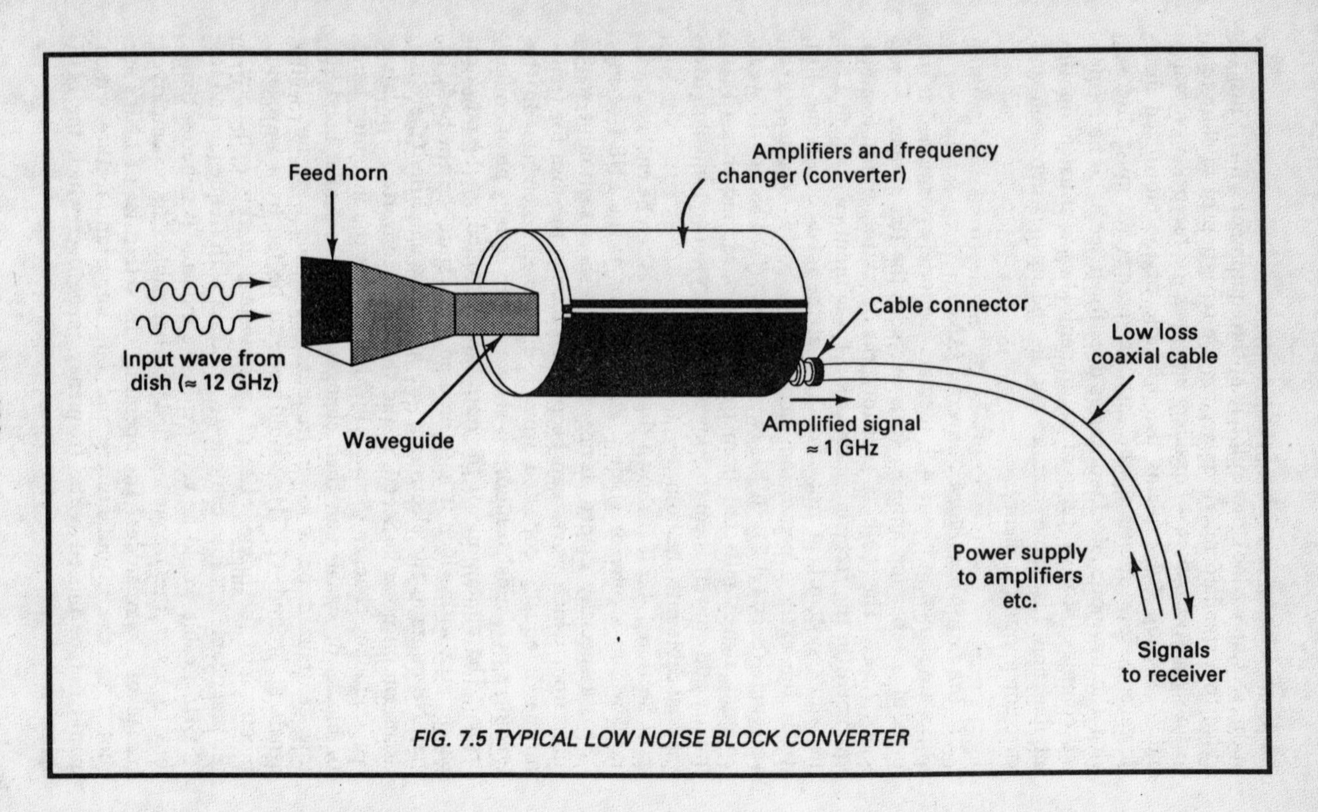

FIG. 7.5 TYPICAL LOW NOISE BLOCK CONVERTER

From Section 4.5 we note that there are four different wave polarizations: vertical, V, horizontal, H, left-hand and right-hand circular, LHC and RHC. The feedhorn of the LNB is sensitive to these and therefore the LNB must be rotated to match the polarization of the incoming wave. Rotation from this position by 90° (a quarter of a turn) will then produce minimum signal pick-up (but maximum for a signal with opposite polarization).

Alternatively feedhorns are available with a *polarizer* or *polarotor* built in. The change from one polarity to the other is effected by rotating a flexible membrane within the waveguide. It is switched into either of the two positions by an electromagnet remotely controlled from the indoor receiver. Nowadays most antennas are fitted with polarotors.

Figure 7.6 gives us an inkling of what goes on inside an LNB. Some typical frequency figures are added for realism. The satellite signals from the feedhorn are first amplified after which they meet a band-pass filter which removes unwanted frequencies outside of the LNB range. The filter is followed by a frequency changer which comprises a local oscillator, mixer and band-pass filter as shown (frequency changing is considered in Sect.5.6). The new carrier frequencies are now 10 GHz lower than those arriving at the antenna, i.e. as shown, within the range 1.2 – 1.45 GHz. Whatever channel frequencies are used, each carries the 27 MHz television band (Sect.5.4). The new band of carrier frequencies is known as the *first intermediate frequency* (1st IF) and is suitable for transmission over the coaxial cable to the satellite receiver indoors.

7.2.2 The Satellite Receiver

This unit as shown in Figure 7.2 is directly coupled to the outside LNB. Receivers now come in all shapes and sizes with so many extra facilities built in that we can do no more than gain a mere conception of what goes on inside. This is sketched in Figure 7.6. The signals arriving from the LNB are still relatively weak and so are amplified first. They are then passed to a system of tuning circuits and frequency changer combined. By controlling together the frequency of the local oscillator and the band-pass filter preceding it, the single channel required is selected and its carrier frequency changed to say, 70 MHz, the 2nd IF. By so doing the channel is "tuned in" and it is again amplified (2nd IF amplifier) and then *demodulated*, the process which regains the original modulation from the final carrier at 70 MHz. The modulation frequencies cover a band containing the video and audio information for driving the television set. Some television sets have special input sockets for video and audio (as already mentioned in Section

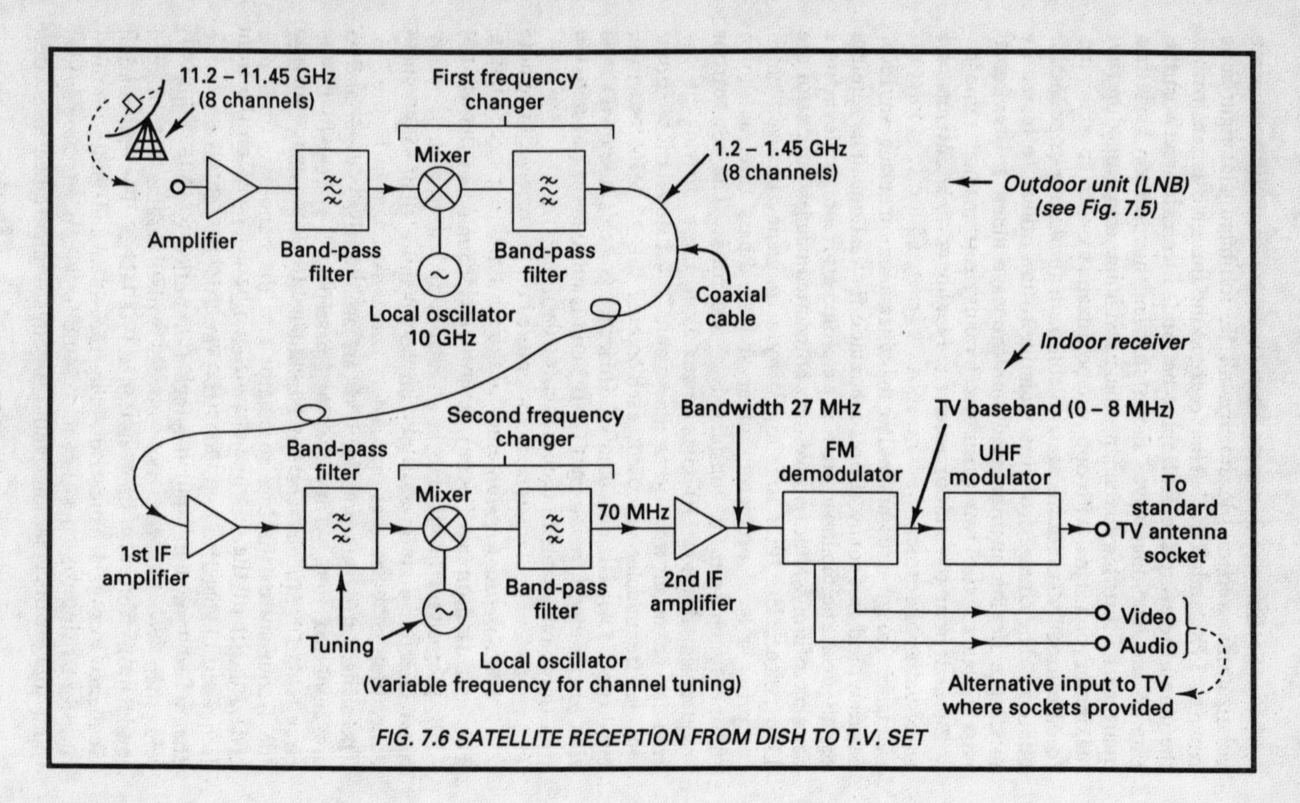

FIG. 7.6 SATELLITE RECEPTION FROM DISH TO T.V. SET

7.1.3) and these are shown on Figure 7.6. When the television set has an antenna input socket only, a UHF modulator is required as shown. In this case the baseband signal modulates a carrier within the normal television set range, this carrier is then fed directly to the antenna socket. Figure 7.6 shows the bare bones only, there are many other features which may be provided as mentioned below.

Many may ask why we do not take the additional step which has revolutionized disc recording (the compact disc), i.e. "go digital". Unfortunately although digital techniques have much to offer, for satellite television each channel would need a much greater bandwidth, in fact several times as much as we use at present.

Without doubt satellite receivers are highly complex devices and are usually microprocessor controlled. As such a number of interesting facilities can be provided to make the complicated business of satellite finding, polarization, polarization offset, channel tuning, etc., less demanding, but at a price. Some of the facilities for a manual or fully automated system with microprocessor control are listed below:

(i) remote control of the receiver – a portable pad with buttons as is used for television (see Fig.7.2);
(ii) the channels may be programmable. All the operations required to select each channel are entered first (e.g. antenna position, channel frequency). This data is held in an electronic memory then brought into action automatically when the required channel number is entered;
(iii) individual parameters can be entered on the receiver or remote control pad (e.g. antenna position, input frequency tuning, polarization and skew) when searching for a channel not programmed as in (ii);
(iv) control of video and audio levels input to the television set for best balance;
(v) the connection sockets on the rear of the satellite receiver may include one for the normal terrestrial television antenna so that the television set can be switched to either source without plug changing;
(vi) details of the required channel can be displayed on the front of the receiver or on the screen of the television set;
(vii) channels considered unsuitable for the kiddies can be "locked" so that they are only accessed by entering a special code;
(viii) buttons or a rotary switch may be there so that the IF bandwidth can be changed (usually reduced). Electrical noise which gets through on a channel is proportional to the bandwidth of the channel. Accordingly if the normal bandwidth of 27 MHz

is restricted, so is the noise. This facility is useful when for example, we are desperate to tune in a channel from a satellite which is not among those producing a reasonable signal in the local area. Because the signal is weak, the total noise arriving may be sufficient to break up or otherwise spoil the picture. By switching to a reduced bandwidth, say as low as 18 MHz, the signal-to-noise ratio is improved accordingly. It is thus possible to end up with a stable picture although some of the detail will have been lost. It is a matter of compromise; we cannot have it both ways;

(ix) an antenna positioner may also be built into the receiver;

(x) even a signal level meter may be added for help when tuning manually.

An old peep-show rhyme says it all for us, "You pays your money and you takes your choice" – but it also depends on how much you pays!

7.3 The Home Installation

For many, choosing their own satellite receiving system may be difficult because naturally one would like access to as many satellites and their channels as possible. However it may not be so easy to gild the lily except at a high price and perhaps ending up with an over-complicated system and quite frequently too large a dish. Remember also that decoding (Sect.5.8) may be required for encrypted channels. For the UK the Videocrypt decoder is at present used but for channels intended for other countries, this decoder may not be suitable so that a different type will be required. Lists of the television programmes for weeks ahead are available from satellite periodicals many of which also contain information on languages, footprints, equipment availability, etc.

Let us take an overview of the types of system which are available and this we do in order of complexity and therefore also of cost.

(i) The simplest comprises a basic fixed dish, LNB and receiver for viewing certain channels on one satellite (or series of satellites) only. This is not as restrictive as it seems, for example all the channels intended for one country only and broadcast from a single satellite can be received on a fixed dish. There is also less to go wrong out of doors than with the systems which follow.

(ii) Given that something better than the fixed dish is required, there are again many options which generally apply to both dish and flat antennas. Other channels on the same satellite but of the opposite polarization can be added. For this it is possible to install two LNB's fitted together in such a way that both types of polarization are received together. However a coaxial line from each to the indoor receiver must be provided, moreover switching between the two can become a little complicated. Alternatively one can visit the dish outside and physically rotate the LNB, a method hardly to be recommended for dishes high up and on wet days. Perhaps the most convenient way is by the addition of a *polarotor.* This is fitted in (or to) the LNB and is remotely controlled from indoors to adjust the polarization, i.e. it is carried out by a flick of a switch rather than by a trip outside. One method of control is by changing the LNB supply voltage from the satellite receiver.

(iii) The range of (ii) can be further increased by more visits to the dish to align it with different satellites. Although all satellites giving a sufficient signal for a particular dish can be brought in by manual adjustment of LNB and/or dish, the method is hardly to be recommended on a regular basis, lining a dish up accurately to a satellite can be a teaser.

(iv) It is also possible to install two LNB's but not for sorting out polarizations as in (ii) but in fact for receiving programmes from two different satellites. This "multi-feed" system works on the principle that although incoming signals to a dish from the satellite to which it is aligned are focused onto the normal LNB situated at the focus (see Figs.4.1 and 4.2), signals from other satellites not too many degrees away, are equally reflected but now focused to a point nearby. This is illustrated pictorially in Figure 7.7 in which satellites 1 and 2 feed LNB's 1 and 2 respectively. There is a price to pay in that satellite 2 signals are not focused accurately and therefore received signal strength is reduced. This loss of signal is counteracted by use of a larger dish say, whereas for (i) a 60 cm dish may be satisfactory, probably a 80 – 100 cm dish would be required for the two-satellite system. Again there is the requirement of a separate coaxial line from each LNB unless switching arrangements protected from the weather can be used. More than two LNB's may be fitted but with the proviso that the dish size should be commensurate with the weakest signal.

Altogether the systems so far considered provide most of us with more than enough additional television programmes, in fact the simplest of all, the basic fixed dish described in (i) is likely to be the most popular. Nevertheless there are those among us who may wish to go that little bit further to the more expensive motorized dish arrangement which does it all for us and reaches many more satellites. Remember however that for the same satellite output power, the farther the satellite is away, the greater is the dish diameter required for a satisfactory signal.

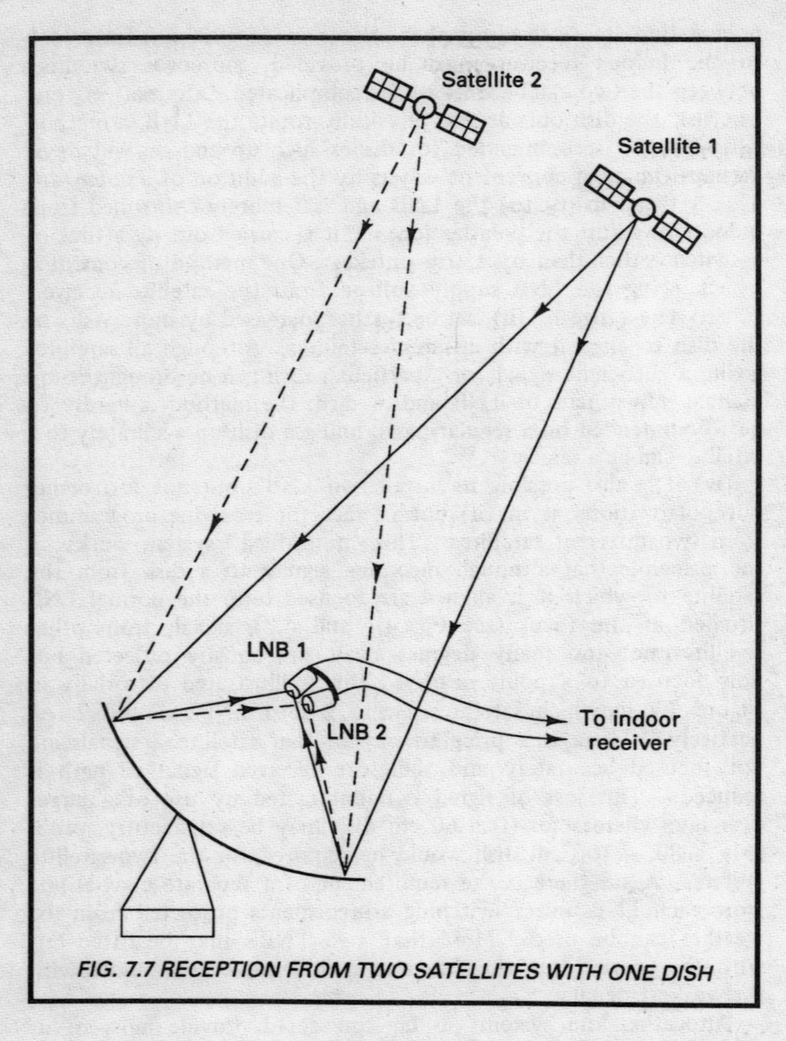

FIG. 7.7 RECEPTION FROM TWO SATELLITES WITH ONE DISH

(v) The motorized dish is introduced in Section 4.8.1.2 (see Fig.4.12 for an example). With this system not only is the dish itself more complicated because it needs a driving system but the indoor control equipment is also. Of course we can dispense with control from indoors if we ourselves go outside to adjust the polar mount manually, an arrangement quite satisfactory for the occasional change. But for the enthusiast who wishes to spend much time searching the skies, a complete system is a must.

The basic principles on which the polar mount is founded are discussed in the earlier Section 4.8.1.2. There is a requirement additional to that of turning the dish which is that of changing the LNB polarization. Either the device can be turned by a remotely controlled motor or what is more likely, a feedhorn can be used with a built-in but remotely controlled polarotor as mentioned in (ii) above (see also Sect.4.5). Generally remote polarization control is automatically provided from the antenna positioner or receiver.

So far we have considered wave polarizations as either vertical or horizontal, the electric and magnetic components being at right angles. It is possible however for a wave to become twisted so that the electric component for example is not truly vertical but at some angle. This is already mentioned in Section 4.5 and is covered in more detail in Appendix 12 which also includes a table of the polarization offsets required for European locations. However we need not be too concerned about this unless we are of the ilk who actually enjoy dabbling in mathematics. Happily this complication is unlikely to affect the home satellite receiver owner because most modern receivers have automatic adjustment built in – it is therefore preferable to ensure that this is so otherwise trips outside may be necessary.

7.3.1. The DIY Approach

There is little doubt that some readers will wish to install as much of the home equipment as possible. There are several reasons, for example:

(i) it is a challenge;
(ii) it leads to a more intimate relationship with the technology;
(iii) maintenance of the outside equipment is in one's own hands;
(iv) there is also the question of cost.

Read this Section thoroughly before making a decision, calling in a dealer when nothing works might be humiliating. Here to help is a collection of reminders, hints and tips. Unfortunately it cannot be a step-by-step guide because installations differ, in some cases widely. Some manufacturers supply complete DIY packages, including tools, instruments and instructions. Even dish antennas can be constructed from DIY kits.

7.3.1.1 Site Survey

First must come a site survey unless it is evident that there is a clear wide view to the South. Section 4.8 covers the principles involved and to make a reliable check a compass for judging azimuth and an inclinometer for elevation are required. These need not be

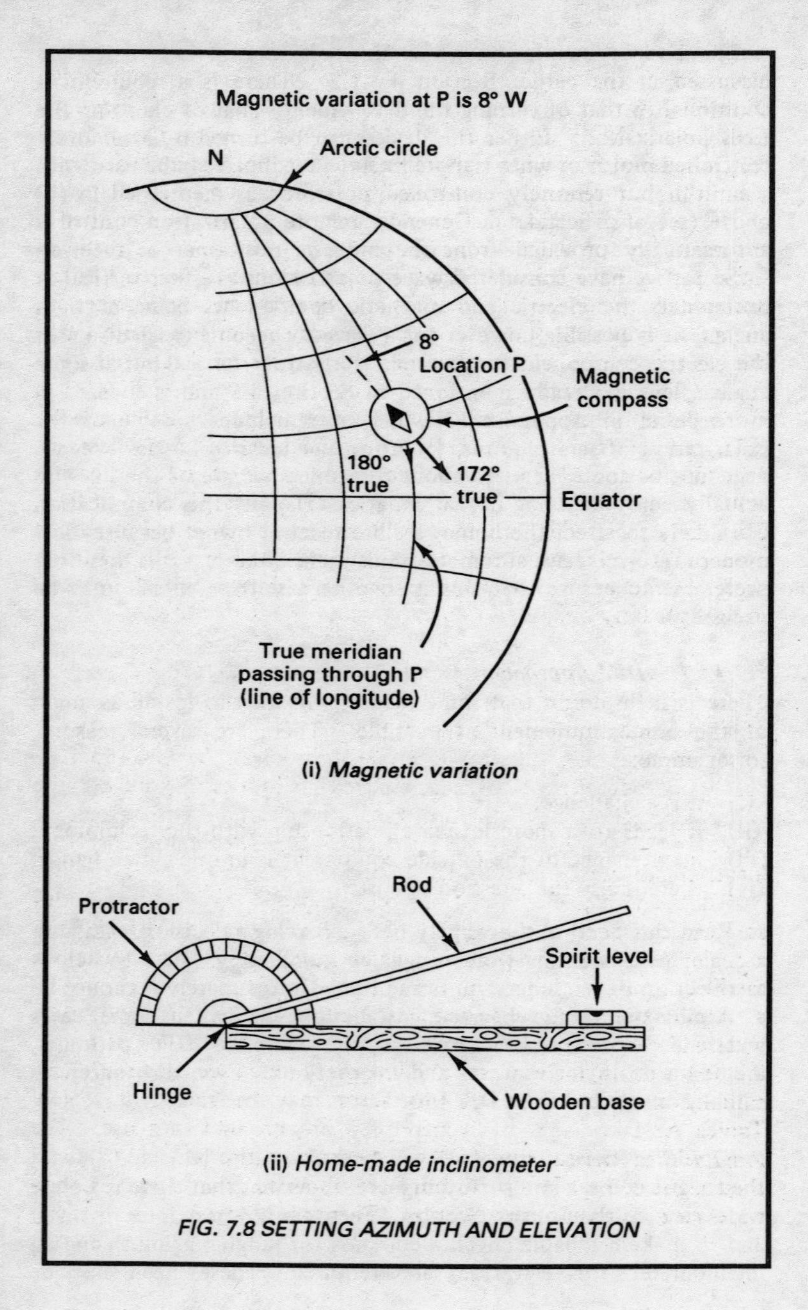

FIG. 7.8 SETTING AZIMUTH AND ELEVATION

complicated devices, for example a pocket compass will suffice but never forget that any compass points to *magnetic* North, not the *true* North we are working with. The present position of magnetic North is in the NW corner of Greenland. What makes life difficult for those concerned with navigation (and for us) is that it does not stay in the same position but slowly wanders around a circular path of about 160 km diameter. From Figure 7.8(i) it can be seen that in the position shown the compass is pointing to the west of North. In some parts of the world it may point to the east of North. In the example the 8 degrees difference is known as the *magnetic variation*, an apt description because it not only varies over the entire Earth but, as mentioned, also with time. It is sometimes referred to as the *magnetic deviation.*

To be helpful, Appendix 8.2 contains the estimated magnetic variations for much of Western Europe on a latitude/longitude basis. The figures are to the nearest degree, greater accuracy is hardly worthwhile considering the other approximations we have to face. They are estimated for January 1996 and can be expected to reduce by about 10 minutes (0.167 degrees) annually. Fortunately for the whole of the area we are considering, the variations are all westerly so we have only one rule to observe:

"**add** westerly variation to true to find magnetic".

This rule is demonstrated in Figure 7.8(i).

As an example, suppose at say, Liverpool in the UK (53.4°N, 3.1°W) an azimuth of 210° is required. From Appendix 8,2 the approximate magnetic variation is 5°, hence the compass reading should be 210° + 5° = 215° (i.e. 215° on the compass at that location indicates 210° from true North).

Although magnetic variations of this order may seem insignificant, we ignore them at our peril for 5° down here translates to only a little less than 5° up there, which could easily be a different satellite.

An inclinometer can be purchased but a wooden base, rod, spirit level and protractor, most of which are to be unearthed at home, are the sole ingredients of a home-made one, see Figure 7.8(ii). Equally a plumb line and protractor may be sufficient. Now Tables A8.1 to A8.4 of Appendix 8 are brought into use. The longitude of the satellite (ϕ_s) will be known, the latitude (θ_r) and the longitude (ϕ_r) for the locality are obtained from a map. Longitudes east are given a negative sign. Then, as shown:

$$\phi_d = \phi_s - \phi_r$$

from which the tables give the azimuth and elevation angles to be used. The figures are given to one decimal place but can be rounded to the nearest degree, our instruments and techniques hardly warrant greater accuracy. The inclinometer is set level and then adjusted for the elevation. Using the compass it is then rotated to the azimuth (plus magnetic variation) so then pointing approximately to the satellite. Repeat for other satellites if required.

If an offset dish is to be used [Fig.4.2(iii) and (iv)], the *dish offset angle* as quoted by the manufacturer must be taken into account. This must be subtracted from the normal elevation figure to give the actual elevation required.

For some of us the result of all this will be that the only antenna position feasible is on the chimney stack and for this there is obviously a limit to dish size. Most dishes, solid ones especially, have to resist the onslaughts of winds so we cannot fit too large a dish to a chimney stack or we might lose chimney and all. The local authority can advise but generally on our homes (but not in the garden) their permission is required for dishes of more than 90 cm diameter. Permission is also required if the proposed dish would project above the ridge of the roof. There are other restrictions referring to flats and houses converted into flats. All this may bring us to a halt with our DIY intentions for the practical difficulties have now increased considerably. Not only does one have to contend with the dangers of ladders and their fixings, working at a height on anything but a flat surface, but also the added complications of lining up to the satellite. Overall, better perhaps to call in the expert installers – don't be venturesome by taking on anything which is at all dangerous.

7.3.1.2 Dish Installation

Once the choice of dish has been made and the goods are to hand, installation begins. We consider a single dish for single satellite working first. It has been suggested that a pointing accuracy of within about half a degree is finally required so care has to be taken. Many different built-in aids are added by manufacturers but if there are none we can employ the makeshift devices suggested above. After all, their help is only needed in locating the satellite, accurate lining up is done afterwards on the television picture or a signal-level meter.

If the dish mounting is a tripod or other metal base, this base must be level. If it is a circular or square tube which is sunk into the ground, this must be vertical which entails checking twice (e.g. with a builder's spirit level), the second time at 90° to the first. We are assuming that the cable run from dish to house is within the generally accepted limit of 80 – 100 m. If greater it may be possible to overcome the excessive cable loss by fitting a special "line amplifier".

To get the azimuth setting correct it may be useful first to indicate the North–South line on the ground by pegs and string. Be careful not to use the compass near to the dish because any iron or steel in the structure will attract the needle and produce a wrong bearing.

7.3.1.3 Lining Up

To help in installation of some systems, their dealers provide installation kits on loan. Such a kit may include compass, inclinometer and even a signal strength meter which can be tuned to the appropriate satellite channel. We assume however that such help is not available.

Undoubtedly the most effective way of finalizing the dish alignment is to have the satellite receiver and a television set connected nearby. This entails making up a coaxial cable (Sect.1.4.5) to connect the LNB to the receiver; the cable from receiver to television set can be the one to be used finally indoors. Don't forget to tune the television set to the r.f. output of the satellite receiver, the latter usually delivers a test signal for this purpose. Mains power has to be run out to the roof or garden for the receiver and television set and great care must be taken, especially in the garden where electric shock can be lethal. On the ground commonsense demands rubber or plastic boots or at least soles of these materials. The golden rule however is never to work on an item of equipment (even indoors) which is plugged into the mains. Cut the power off preferably by pulling out the plug rather than merely switching off. Remember that there are fatalities every year from accidental contact with the mains voltage. Voltages of much less than 100 have been known to kill!

The real danger arises when the human skin resistance is lowered through dampness or perspiration and if we have not pulled out the mains plug as recommended above, then knowing the risk we are taking, we are likely to perspire more. Fortunately homes and individual sockets are becoming safer now that the *residual current circuit breaker* has arrived. This device is inserted in the socket-outlet or in the mains supply to a group of sockets and switches off the supply within about one-thirtieth of a second if slightly more current (a mere 10 – 30 mA) is flowing in the live wire to earth than is flowing in the neutral wire. The supply is therefore disconnected before it can affect a person's heart. For a little extra cost and care therefore we can always be safe.

There is one alternative which is that a member of the family or a friend may be co-opted to watch the television set inside the house and act as a human signal-level meter by calling out the state of the picture to the person struggling with the dish. No equipment is moved by this method but too much guessing has to go on for it to be recommended although it is well known that many terrestrial television antennas are adjusted in this way.

Ensure that the LNB is set for the appropriate polarization and offset (Appendix 12), is pointing correctly and that the receiver is tuned to the desired satellite channel known to be on the air, then switch it all on. There could be a picture but don't bank on it! If only a screenful of white flashes, hold the dish from behind and make a square search of the sky until the satellite is found. This is done by moving the dish to pick out a tiny square, then on the same centre, a second one slightly larger, then again, increasing the square size until the satellite is eventually found. It will first be recognized by a slight break-up of the noise flashes on the television screen. Once this happens, further fine adjustment of azimuth, elevation, LNB and skew control should bring in the picture. Adjust for the best possible. A useful method when the picture is held is to swing the dish slightly to the left, just into noise, then over to the right again just into noise. The best picture should be about half-way between the two positions. This adjusts for azimuth. Repeat up and down for elevation. Finally tighten all adjusting nuts.

Many receivers include a signal level (or strength) meter. Using this when making adjustments for signal level has several advantages over watching the television picture. A good picture may not necessarily mean that the signal is at its best and we must aim for this to have something in hand for the inevitable rainy day (Sect. 5.2.1).

As an alternative some satellite receivers have an *automatic gain control* (a.g.c.) terminal at the back. A.G.C. is an electronic system which maintains the output of a receiver fairly constant irrespective of variations in the input signal level. The signal at the output is monitored and a control voltage commensurate with its level is fed back over the a.g.c. line to the input amplifiers. Here a high level of a.g.c. voltage reduces the amplification and vice versa. The a.g.c. terminal is connected to the a.g.c. line so connecting an ordinary voltmeter between the terminal and the ground (or earth) terminal provides a visual signal level check. The voltmeter can probably be used at some distance from the satellite receiver over a length of twin bell or telephone wire. It is worth trying out as this method may avoid connecting the whole affair up in the garden. Visits indoors or the help of an accomplice will still be required to check the picture but the ultimate fine tuning is carried out more easily with the voltmeter near at hand outside.

The flat antenna should present fewer problems. Typically the plate might be only some 2 cm thick and have a simple bracket mounting at the rear. Being much lighter than an equivalent dish, it can be mounted practically anywhere where there is the mandatory uninterrupted view of the satellite(s), even on a wall or sloping roof. Connection from the LNB to the indoor unit is the same as for a dish.

7.3.1.4 Polar Mounts

An introduction to the polar mount is given in Section 4.8.1.2. This type has appeal to all who wish to change between satellites frequently and it is essential for those wanting to do this from the armchair for it is the only form of mounting which lends itself to being "motorized". The shape of the polar curve has been built into the mount by the manufacturer, setting up involves aiming the dish accurately at the *pole*. The pole is at the highest point of the geostationary arc and because the arc is symmetrical about its pole, for any location in Europe the pole must be at true South (see Fig. 4.11). As might be expected the elevation of the pole varies with the latitude of the dish location.

Tables A8.3 and A8.4 in Appendix 8 can be used to find the correct elevation for a dish pointing South for then the longitude difference, ϕ_d is zero. For example at latitude 40°N (central Spain) the elevation required is 43.7° (Table A8.3) and the dish can be set to this. There is another method however which achieves the same result but with a less complicated table although it hides under a rather dissuasive title "Dish Declination Offset Angles". The full reasoning is given in Appendix 8 (Sect.A8.3) but briefly all that has to be done is to add the appropriate offset angle to the latitude to obtain the declination angle. Table A8.6 lists the standard offset angles for Europe. For the case quoted above (40° latitude) this one is 6.3° hence the declination angle is $40 + 6.3 = 46.3°$. Most polar mounts are conveniently calibrated so that this angle can be easily set.

Figure 7.9 shows how the declination angle relates to the normal elevation angle, i.e. it is converted by subtracting it from 90° so we get $90 - 46.3 = 43.7°$ as above. Summing up:

$$\text{latitude} + \text{offset angle (from Table A8.6)} = \text{declination angle}$$

also:

$$\text{elevation} + \text{declination} = 90° .$$

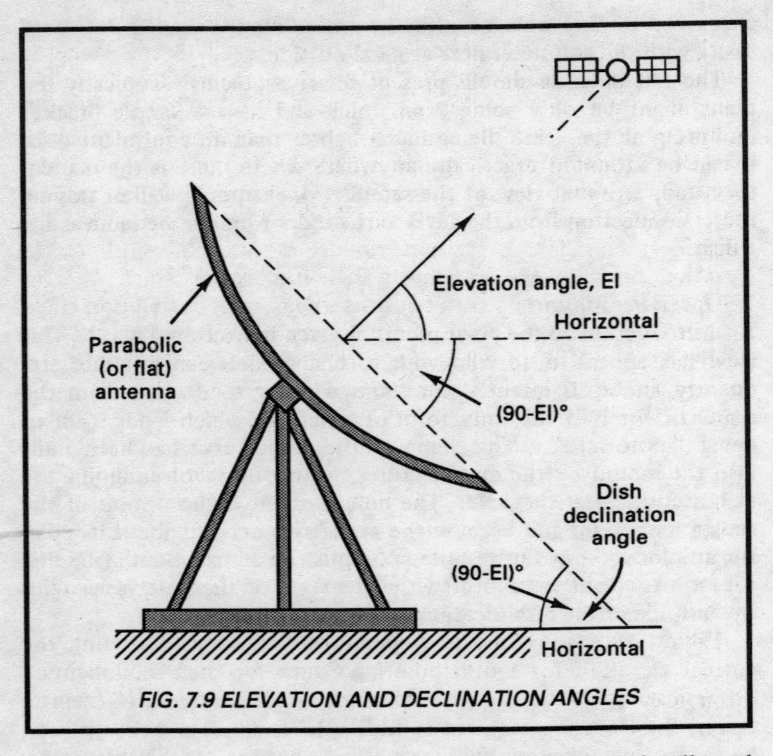

FIG. 7.9 ELEVATION AND DECLINATION ANGLES

A polar mount dish facing true South and so aligned will track the whole arc from horizon to horizon as it is turned. This can be done by hand or motor. If such a dish is not accurately aligned or its support is not truly vertical then the tracking curve will not coincide with the geostationary arc, the result being a wonderfully clear view of nothing.

Although only parabolic dishes have been considered above, the technique applies equally to flat antennas (Sect.4.2.3).

Much of the basic installation work is covered in the preceding sections but there are some differences with the polar mount. For example, as shown, elevation need not be adjusted in the same way and in fact Tables A8.1 – A8.4 are not needed. The only information required is the latitude and offset angle.

If we understand the principles involved then the installation work follows ordinary commonsense rules. Nevertheless, a few additional hints may help.

Firstly the post on which the dish is mounted must be absolutely vertical. To check this a builder's spirit level is useful and the post must be checked twice, 90° apart. The polar axis of the dish must be aligned true N–S, the stretched string pegged to two stakes in the ground may help (Sect.7.3.1.2). Don't forget to add in the magnetic variation when using a compass (Sect.7.3.1.1). Latitude and offset angles are easily adjusted irrespective of how the mount is marked. Elevation or declination angles can be checked by running a taut string from top to bottom of the dish and with an inclinometer measuring the angle the string makes with the horizontal (see Fig.7.9). Remember it is quite in order to set the dish to the elevation instead of the declination angle if more convenient.

Once the dish points true South at the correct declination angle, it is in fact lined up to the geostationary arc. Swinging the dish slowly round to left or right ought then to pick up a satellite near the top of the arc. When one is found make adjustments as necessary to the mount for maximum signal as in Section 7.3.1.3. Then swing to the other side of the arc to pick up a second satellite and again adjust. By swinging between the two and continuing to make fine adjustments each time, the highest part of the arc should be traced satisfactorily. Then move on further to other satellites. If, when in use the polar mount is to be manually adjusted, reference marks made somewhere on the mechanism can help to ensure quick returns. When all is finished, check the bolts for tightness.

Overall hardly child's play and possibly time-consuming but rewarding. What is more, now having an in-depth and practical understanding of the whole arrangement, subsequent maintenance becomes that much more manageable.

7.3.1.5 Cabling

All the bits and pieces have to be connected together and a length of cheap bell wire just will not do. Terrestrial television proves this for all its antenna downleads comprise coaxial cables (Sect.1.4.5). As a reminder, one conductor is the outer tube, the other is a wire running centrally through the tube (see Fig.1.3(i)) and kept in position by a special foam or solid insulating discs. Such cables require special connectors. When the signal travels along a coaxial cable there is the inevitable loss (it would soon be lost altogether over a length of bell wire). Generally for the same materials and method of construction, the larger the diameter of the cable, the lower is its loss. Typically a 10 mm diameter cable might have a loss of 12 dB per 100 metres, whereas at 5 mm the loss is more than double.

Although both signal and noise suffer the same loss when travelling along a cable and on this account the signal-to-noise ratio (Sect. 1.5) does not change, there could be the problem of electrical noise being picked up by the cable itself. Coaxial cable has the advantage of high immunity from interfering signals and noise from outside due to its special construction. Nevertheless it cannot be completely immune hence it is important to aim for low loss and low pick-up which means installing the dish as near to the house as possible. It would be unwise therefore to install a dish out of sight at the bottom of a long garden using a cable appreciably more than say, 50 – 60 m long (unless a special amplifier is added). Avoid nasty bends or kinks in multi-cables, at least one wire is feeding power to the LNB so internal short-circuits may create havoc and because they are internal, will probably be difficult to trace.

Because the fully automated dish (remote control of polarizer and actuator) needs extra wires for control, special multi-cables are available consisting of the signal coaxial cable with one or more non-coaxial cables combined.

7.3.1.6 Indoors

A brief discussion on the indoor satellite receiver has already been given in Section 7.2.2 with Figure 7.6 showing the main features from dish to television set. We must leave it at that for so much depends on individual choice and the funds available. The total number of satellite television receivers in the world moves inexorably on and it *might* even be predicted that sometime in the future satellites will take over from terrestrial systems altogether.

Chapter 8
OTHER SERVICES

Chapter 7 concentrates on the use of satellites specifically for the purpose of distributing television programmes. To most of us this is what the term "satellite" engenders. However there are many satellites quietly doing a job of perhaps greater importance such as improving safety at sea and in the air, helping our armed forces, looking down from above to assist in weather predictions and even in co-operating with international spying. We discuss some of these facets of satellite use in this chapter, many of them would require a whole book for a complete examination. Here of course we can do no more than to mention briefly some of the major features.

8.1 Mobile Systems

Communications with ships at sea had its early beginning around 1900, mainly for help in emergencies. Systems of those days were based on telegraphy and used the Morse Code, hence the distress signal, SOS. By the early 1920's two-way voice communication at HF (Appendix 3) had arrived and the distress signal changed to "Mayday" (representing the pronunciation of the French "m'aidez" – help me). However the HF band suffers in its propagation through the ionosphere and does not always provide a stable signal, moreover the bandwidth available is limited. As soon as radio transmission over higher frequencies (VHF) became feasible, telephony transmission to ships was developed but this was only possible to vessels within a few tens of kilometres of the coast. The advent of satellite systems has therefore been of inestimable benefit.

INMARSAT (The International Maritime Satellite Organization) is an example of a system which arose mainly because of the needs of shipping for a global maritime distress and safety system but it has now expanded to provide communication services not only for shipping but also for aircraft, off-shore industries and land-based users. There are at present 65 member countries involved, all of which share in the policy-making and financing of the system. Working frequency bands are 1.5 – 1.6 GHz and 4 – 6 GHz.

The main facilities of INMARSAT are provided by three operational satellites in geostationary orbits directly above the Atlantic, Pacific and Indian Oceans. They are effective up to about 80° latitutde although at or approaching this angle, transmission problems may begin to arise.

8.1.1 Land Mobile Systems

Interest in land mobile satellite services arose as soon as it was possible to exploit the developing technology in the early 1980's. Since then development has proceeded apace, accelerated by the need to provide coverage of large country areas in which few or no communication facilities existed. As an example such coverage is especially useful to the international road transport industry.

When the position of a satellite receiving station is fixed, practically no problems arise due to signal reflections from nearby objects. When the station is on the move however it invariably has a low gain antenna and signal reflections from elsewhere are added to the required signal. Each reflection has travelled over a longer path and therefore arrives differing in both amplitude and phase. The result is that the instantaneous received signal level varies with time as the vehicle passes through an interference pattern. Fading of the signal is also experienced when a vehicle is shielded from the satellite by buildings, other larger vehicles, trees etc. Obviously fade durations are dependent on vehicle speed and whether the journey is being made in town or country. Complicated electronic techniques are therefore built into the system to minimize these effects.

8.1.2 At Sea

(1) for communication, a complete connection from land to a ship at sea over the INMARSAT network involves:

(i) coast earth stations (CES) connected to the national telephone network with large antennas (10 or more metres diameter) transmitting and receiving to and from the satellite;

(ii) the satellite itself, acting simply as a relay station but with frequency changing built in;

(iii) a ship earth station (SES) having a much smaller antenna than the CES (e.g. 1 – 2 metres). It goes without saying that facilities must be provided on board for continually steering the antenna and to compensate for roll of the vessel, e.g. by gyroscope motors. The above-deck equipment (antenna, positioner, transmitting amplifier and LNB) is contained within a glass-fibre *radome* for protection. Below deck, computer control is most likely to be used, driven by a crystal-controlled oscillator. This ensures that accurate intermediate frequencies are generated, leading to a final equipment output of for example, audio or PSK [phase-shift keying – see Fig.5.3(iii)] ;

(iv) for navigation – accurate world-wide positional information is now provided by special satellite systems, one of which is the *Global Positioning System* (GPS). This employs as many as 24 satellites spread around three circular orbits some 20 000 km high resulting in a 12 hour period. Such an arrangement ensures that at least six of the satellites are in radio contact with any location on the earth at any one time. Transmissions from the satellites enable a ship to fix its position by measurement of the distance to each of three satellites. An accuracy of position to about 100 – 150 m is normal but higher precision is available if required, to as little as a few centimetres or for scientific applications, even to millimetres. Naturally such precision requires special techniques and computation methods. Since first development the system has been extensively adapted to many civilian applications especially in off-shore surveying and in the study of the distribution of minerals (such as oil) on the earth. There are even hand-held GPS receivers enabling users to obtain their latitude and longitude readings to about 200 m accuracy "at the press of a button".

8.1.3 In The Air

Terrestrial systems, i.e. using direct air-to-ground links are already well established and are especially useful where much air traffic flows mainly over land. On the other hand satellite based systems have many advantages when coverage is required for aircraft flying over transoceanic routes. Unfortunately this type of use has had to await the development of sufficiently light and compact equipment for the aircraft but this has now arrived. Present satellite communication systems for aircraft use the INMARSAT network and provide data, facsimile and normal voice telephony for passengers. Perhaps of even more importance is the fact that operation and control information can also be transmitted to and from the aircraft so improving flight efficiency especially over the longer transoceanic routes.

8.1.4 News Services

When news is carried by speech only, the normal land lines are ideal; moreover, connections to most locations are set up easily. Television however requires considerably more than a thousand times the bandwidth of a speech channel hence on earth special wide-band circuits must be connected up in advance as for example, for major events such as football matches and other sporting activities. Satellites can handle this type of traffic also. However when it comes to events which arise at short notice and can be anywhere in the world, there is little time to set up land connections and in fact, for many events in the wilds such connections are likely to be unavailable. Film cameras are not considered in this context because of the delay in processing. This is where satellites come into their own because arrangements can be made for portable equipment to be deployed rapidly for transmitting over an uplink from most places on earth. The system is known as *Satellite News Gathering* (SNG). The fact that the equipment is portable means that it can be shifted quickly to other events. Portability however is not as we would normally consider it. The heaviest items are the antenna which is likely to be 1.0 – 1.5 m in diameter and the power supply which, whether by battery or generator, has to provide more than, say 300 watts, most of which is taken by the high-power transmitting amplifier. At least two people are therefore required for transporting a complete system.

Present equipments work mainly in the 12 – 14 GHz frequency range. Generally to ensure successful operation, practical satellite news gathering systems transmit up to the satellite at an e.i.r.p. of 65 – 70 dBW (Sect.5.2). The system also processes downlink signals especially for initial antenna alignment and for monitoring.

8.2 Data Broadcasting

There is a system whereby information in digital form (the data) is delivered to several receiving ground antennas. The use and desirability of such a system can be appreciated by considering organizations such as airlines, oil companies and departmental store chains which need to send out information from say, a head office to the units such as branch offices, stores, etc., which it controls and which may be spread over a wide area. Such facilities already exist over the normal telephone network but the cost of private lines to each of the subsidiaries may be much greater. A simple data broadcast system is illustrated in Figure 8.1, it is generally classified as "point-to-multipoint".

The data transmitted is in the form of "packets" or blocks of bits. Each packet can be likened to a letter sent through the post, consisting mainly of the data itself and the destination address. In addition start bits are required for synchronization and the message itself may contain additional bits for error protection. If the full message occupies more than a single packet, a packet sequence number is included. The originating address may also be required and finally an end bit sequence signals the completion of the packet. The packet address ensures that only authorized users receive the information unless as for example with press releases, all users require access. Each receiver can therefore identify the packets which are addressed to it and extract the data from each packet, this is then transmitted to the user.

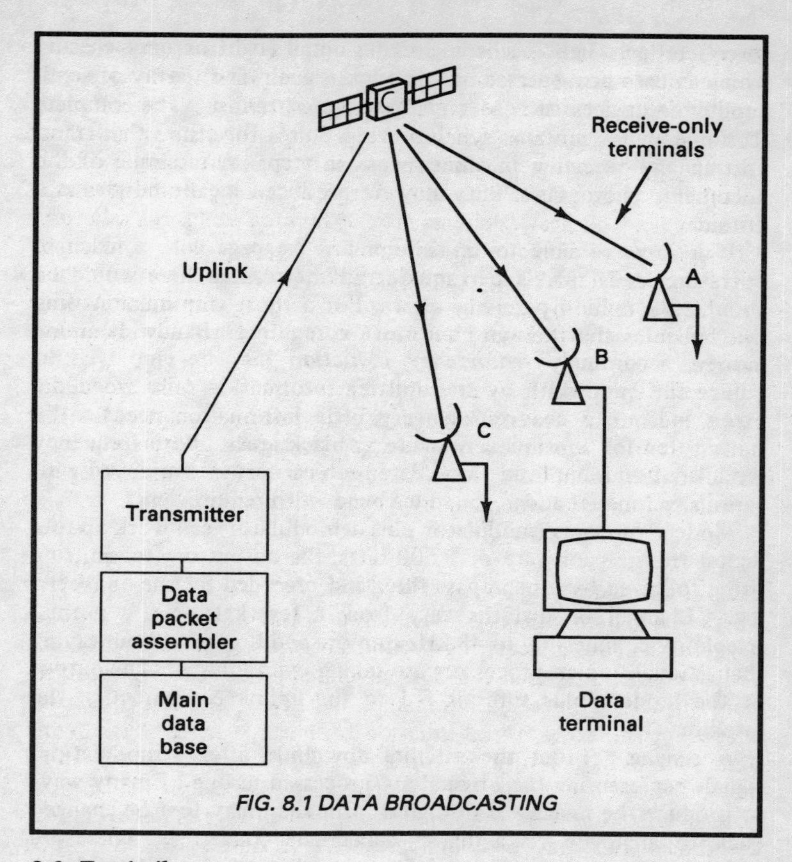

FIG. 8.1 DATA BROADCASTING

8.3 Facsimile

This is the process of making a copy of writing, printing, photographs, etc. It was originally developed for transmission over telephone cables but has more recently become available over radio systems and especially over satellite links with their wider reach. Systems have now reached general transmission times of some three minutes for a single A4 document (210 × 297 mm) but down to one minute for a high speed service, even a whole newspaper in less than one hour. Analogue systems cater especially for pictorial transmission whereas digital systems function better where only black and white is used, e.g. printed documents.

Transmitting. Naturally the system has some parallels with television in that scanning is involved. A document is scanned sequentially by a light beam crossing from one side to the other, at the same time moving from top to bottom. Each picture element (*pixel*) reflects light according to its detail (light or dark colour) which is used to generate an electrical signal via an array of semiconductor devices such as a *charge-coupled sensor.* The complete facsimile signal contains synchronizing pulses to ensure that transmitting and receiving machines work in step. A facsimile of the document, photograph, etc., may be produced locally besides at a distance.

It is now possible to transmit a full A4 page with a machine operating at 30 pixels per square millimetre, i.e. there would be about 1.87 million pixels in total. For a short transmission time this indicates that a large bandwidth is required. Bandwidth costs money, accordingly *redundancy reduction* may be employed to reduce the bandwidth by transmitting information only when the pixels indicate a change, e.g. very little information needs to be transmitted for continuously white or black areas. Both frequency modulated and amplitude modulated sub-carriers are employed with various techniques added to reduce bandwidth requirements.

Modern modems (modulator plus demodulator) can work up to a digital transmission rate of 9 600 bit/s, the output on transmitting being followed by a band-pass filter and preceded by one on receiving. Channel bandwidths vary from a few kHz (e.g. a normal telephony channel) up to 48 kHz or more and it goes without saying that gradually microprocessors are taking over control. The output of the modem plus filter is fed to the uplink equipment of the satellite.

Receiving. From the satellite downlink, after demodulation, signals representing the original are processed in one of many ways to produce the image. If in digital form they may first be changed back to analogue via a digital-to-analogue converter. There are many systems in use for producing a replica of the original document, photograph, etc.; too many to describe in detail here. They include:

(i) *electrolytic:* a special conducting paper is used. A scanning stylus passes current through the paper to a conducting plate behind. The paper then darkens according to the current strength;

(ii) *electrostatic:* a special electrically charged stylus is made to scan the paper, leaving a charge pattern on it. The paper then passes over magnetized iron particles mixed with a black iron powder. The powder is attracted to the charged areas according to the magnitude of the charge and so builds up the image which is bonded permanently to the paper by a subsequent heating process;

(iii) *thermal:* the method is in common use but requires a special thermosensitive paper which discolours according to the temperature to which it is subjected. Heating is via an array of minute resistors produced by thin-film technology, such resistors heat up or cool very quickly and therefore are able to respond to a varying input signal;

(iv) *photographic:* a photosensitive paper is scanning by light, for example, from a laser driven by the incoming signal. Alternatively the light may be fed by optical fibres to the surface of the paper.

These are only a few of the modern techniques employed in facsimile printing, cathode-ray tubes and light-emitting diodes in conjunction with a fibre-optics array are also employed; much depends on the final image required.

Undoubtedly with further development and with the increasing use of satellite and land-based long-distance circuits, facsimile transmission will become increasingly competitive with existing postal services.

8.3.1 Weather

There can be very few people who have yet to experience the enchantment of television weather reports in which we are actually able to look *down* on our weather and even watch the banks of clouds as they obscure parts of the country. It all looks so simple but behind it is a complex system of geostationary and polar orbiting satellites working together to provide the earth not only with pictures from up above but also with forecasts for shipping, aircraft and many other operations which rely on knowing what weather conditions are on the way. The geostationary satellites are at present the joint efforts of the USA, Russia, European Space Agency and Japan. Polar orbiting satellites are operated by the USA and Russia. All told, these satellites provide weather services for most of the earth and therefore are able to sense weather patterns such as storms as soon as they begin to develop.

The geostationary satellites are spaced around the equator starting at 0° and are about 70° apart and each covers a latitude up to some 75°. The areas not covered by the geostationary satellites are handled by the polar orbiters. Each satellite carries a *radiometer*, a device for measuring the radiant energy from earth. The satellite itself has an axial spin and its radiometer scans the earth from east to west. After each scan the radiometer is moved slightly in a north–south direction so that a full image is built up in about 30 minutes. The radiometer output is then processed to provide information which is signalled to earth, typically as follows:

(i) *visual images* taken in daylight hours and showing the earth below;
(ii) *infra-red radiation* which is proportional to the earth temperature and can be assessed during both day and night. Space is cold hence appears on the image as white whereas hot areas of the earth are dark. An infra-red image therefore enables temperature changes on earth to be assessed;
(iii) *water vapour* which enables the humidity in the upper atmosphere to be monitored.

The information is transmitted in digital form to the main earth meteorological stations, i.e. the *primary data users* and in analogue form to *secondary data users* on frequencies between 1.6 and 1.7 GHz. The primary users have large antennas (4 m or more diameter) and receive higher definition images. They carry out further processing before sending the data onwards to other users.

Secondary data users require smaller antennas, they also receive images direct from the satellite, these results are more easily processed and displayed by cathode-ray tube or even by facsimile machines. Meteorological centres naturally employ more complex processing techniques although at the other end of the scale there are "home" installations providing the more interested viewers with facilities for studying weather charts immediately they are produced.

8.4 Military

For such purposes the range of SKYNET satellites carries much of the traffic, these are generally described as geostationary defence communication satellites, providing facilities for all three services, i.e. army, navy and air force. Communication via these satellites is particularly attractive because they provide a stable service over a large coverage area. At the high satellite frequencies, each system has high capacity, especially compared with that available from HF systems, moreover there is not the problem of erratic propagation conditions. A main disadvantage is that such links may also be available to others for whom they are not intended (e.g. an enemy) hence back-up, non-satellite systems are used in addition.

Antennas on geostationary military satellites tend to be designed for wide coverage since the communication requirements may be unpredictable in their location. The more recent ones may be operational at both UHF and SHF (Appendix 3) with the antennas selected as required. At UHF the transponders are likely to be solid-state, delivering perhaps some 40 watts of r.f. power whereas at SHF only travelling-wave amplifiers are suitable.

Army systems frequently work to small land-mobile terminals. The mobility required may be high under battlefield conditions. The satellite may transmit data, computer control signals or speech. The latter is usually in digital form and is most likely to be encrypted. The smallest land terminal is the highly portable "man-pack", a complete terminal which can be carried by one man. The dish size is some 45 cm and it must be aligned accurately to the satellite. Because of the small size of the terminal it performs best in regions covered by a spot-beam antenna.

Ships have additional problems to overcome, movement of the vessel makes antenna pointing complicated and at times parts of the ship's superstructure may block the transmission path. This latter difficulty may be alleviated by the use of antennas on opposite sides of the superstructure. Use of the system by submarines is only at UHF with a small antenna projecting above the water surface.

For aircraft the problems multiply mainly owing to the difficulty of antenna pointing and stabilization, especially for high-speed fighter aircraft. Size of antennas is also severely restricted.

Needless to say there is always the enemy to consider. Ground stations are of course vulnerable so duplication may be essential. Much of the station equipment may be contained underground but unfortunately its dish antennas must be on the surface. At present the satellite itself seems not to be under threat although no doubt research is going on somewhere with a view to destruction of satellites by high-energy laser or particle beams.

Jamming by an enemy is accomplished by swamping either the uplink or downlink with high power. Uplink jamming can run the satellite transponder into overload so that the downlink signal-to-noise ratio is greatly reduced. Downlink jamming on its own is less of a problem since at present it can only be accomplished by an aircraft which itself is vulnerable. Difficult though the problem may be, many techniques exist which while not completely neutralizing the jamming, certainly make it less effective.

Chapter 9

THE FUTURE ?

Telephony and telegraphy have been with us for around 100 years. Then along came facsimile which enables still pictures to be transmitted followed by television by which live pictures are transmitted. When television first entered our lives however transmission across the oceans was a goal then unattainable. Although undersea cables offered some hope, the consumption of bandwidth by a television programme was dissuasive. However the development of artificial satellites orbiting near the earth followed by those in the geostationary orbit came to our rescue. Television and in fact everything else by satellite was with us. Satellites have got bigger whereas their earth stations have become smaller.

Lest we now think that satellites are the be-all and end-all of world-wide communication, there is now competition from the optical fibre which can carry television programmes and the like by cable across oceans with ease. Even on land the fibre rivals the traditional but expensive coaxial cable and the microwave link with its limited frequency range. In addition fibre systems suffer less from the problem of transmission time which as we have seen detracts from the excellence of satellites and cannot be lessened. Nevertheless optical fibre cables must be laid in the land or sea whereas satellite transmission enjoys the freedom of the air.

At present most satellites are merely relays in that they simply receive information and then pass it on. In their passage through the satellite the signals are amplified but not significantly changed. It is possible that the future will bring "intelligent" satellites, i.e. those which process the incoming signals in some way or even switch them to different destinations as we now do in telephone systems. Business networks may develop rapidly especially for systems which employ one large earth transmitting station "broadcasting" information to hundreds or perhaps many thousands of small inexpensive receiving terminals. There is also the need for these terminals to reply and for the expansion of small mobile stations.

Much of our discussion has been in terms of geostationary satellites, however it does seem that low earth orbits, until now rather difficult to handle, may becoming back into fashion. Although they have their tracking difficulties, they possess several advantages over the geostationary type especially in the use of a shorter transmission path with, say a mere 40 seconds delay compared with the geostationary at 240 ms (Sect.6.1). The satellite-to-ground loss is also much less by as much as 20 dB hence accurate antenna alignment is less of a problem. However complications arise especially in that as viewed from the ground, each satellite rises from one horizon and disappears over the other within a fraction of an hour, accordingly several satellites are required per system. However they are much smaller than the geostationary ones and considerably less costly. Highly elliptical orbits may also be used with perigees of less than 500 km but with apogees up to tens of thousands of kilometres, giving much longer communication time per orbit. Altogether it seems to be a reversal of present trends but in fact merely complementing them.

Nowadays satellite beams are changed by switching to a different antenna specially designed for the particular purpose. Present day research aims to produce antennas which can change their beam configurations automatically or on instructions from the ground. This technique will increase the flexibility of the satellite and at the same time reduce the number of its transmitting antennas.

It would appear that, far from being eclipsed by the optical fibre, satellite transmission which has already spread widely into the television distribution and many other fields has much yet to offer, especially in its flexibility and versatility. We await this eagerly, perhaps it is a case of more of the same but even better!

Appendix 1

GLOSSARY OF SATELLITE COMMUNICATION TERMS

This is a list of terms and abbreviations readers may find here and in other satellite literature. To the experienced engineer some of the explanations may appear imprecise and even naive, but this is deliberate in an attempt to help those less technically minded.

ABSOLUTE TEMPERATURE – is the temperature measured from absolute zero (–273.16°Celsius).

ABSORPTION – the reduction in intensity of an electromagnetic wave by a material.

ACTIVE SATELLITE – a satellite capable of receiving, amplifying, processing and then re-transmitting electromagnetic waves.

ACTUATOR – is the motor and its feedback system used to drive a polar mount antenna.

AERIAL – see Antenna.

ALTERNATING CURRENT (A.C.) – is an electric current which reverses its direction of flow at regular intervals. Radio waves produce alternating currents in antennas, speech and music produce alternating currents in a microphone. The electricity mains supplies alternating current at 50 Hertz (cycles per second).

AMPLITUDE – the strength or magnitude of a signal.

AMPLITUDE MODULATION (A.M.) – a method of impressing a signal on a carrier wave by varying its amplitude (Sect.5.4).

ANALOGUE – this usually refers to a mode of transmission of information. An analogue waveform has a physical similarity with the quantity it represents and therefore can usually be expressed by a graph on a base of time. Typical examples are given by the output of a microphone, television camera or any device measuring a quantity which varies with time such as temperature, pressure, etc.

ANTENNA – a device used to transmit or receive radio waves. Those used with satellites work at very high frequencies and are usually of the parabolic dish type. In the UK the term "aerial" has been and still is used.

APERTURE – as applied to an antenna is the area from which it radiates or receives energy.

APOGEE – the point farthest from earth in the orbit of, for example, the moon, a planet or an artificial satellite. A geostationary orbit is circular and therefore has no apogee.

APOGEE KICK MOTOR (AKM) – a rocket motor installed in a satellite which moves the satellite into its final orbit (Sect.2.1.1).

ARIANE – the European expendable launch vehicle which puts satellites into orbit from the base at Kourou, French Guiana.

A.R.Q. (AUTOMATIC TRANSMISSION REQUEST) – used with data systems. Transmission is repeated if the receiving end detects an error.

ASTRA – the general name for a group of 16-channel medium power satellites situated at 19.2° East and operated by the Société Européenne des Satellites (Luxembourg).

ATMOSPHERE – the gaseous envelope surrounding the earth.

ATMOSPHERIC LOSS – loss suffered by a radio wave in travelling through the earth's atmosphere, usually quoted in decibels (Sect. 5.2.1).

ATTENUATION – the reduction in amplitude of a signal through power losses in the channel over which it is travelling.

ATTITUDE – the position of a spacecraft or satellite relative to specified directions.

AUDIO – that which we can hear. The term is also used to describe the electrical representation of speech and music.

AUDIO FREQUENCY – any frequency of a sound wave which can normally be heard. The maximum range is from about 20 Hz to 20 kHz.

AUDIO SUBCARRIER – a carrier frequency above the main vision carrier in a television signal providing the main audio channel (i.e. the one for the picture) plus as many as 8 extra audio channels.

AUTOMATIC FREQUENCY CONTROL (A.F.C.) – a circuit in a radio or television receiver which ensures that the tuning circuits remain correctly adjusted to the incoming wave frequency.

AUTOMATIC GAIN CONTROL (A.G.C.) – a circuit in a radio or television receiver which maintains the output of an amplifying stage relatively constant irrespective of variations of the signal applied to the input (also known as "automatic volume control").

AV – stands for audio/visual; used where equipment caters for both.

AZ/EL MOUNT – (Azimuth/Elevation). The basic parabolic antenna mount. Both azimuth and elevation are adjusted separately.

AZIMUTH – the horizontal angle measured from true North to a line joining an observer to a satellite.

BAND-PASS FILTER – an electronic circuit which passes a predetermined band of frequencies only. It does so by presenting high attenuation to all frequencies above and below the band, thereby preventing them from reaching the output terminals.

BANDWIDTH – the range between the highest and lowest frequencies in a communication channel, measured in hertz.

BASEBAND – the range of frequencies initially generated and which is subsequently transmitted by radio (e.g. from audio or television studios).

BASIC – (Beginner's All-purpose Symbolic Instruction Code). A popular computer language, i.e. the series of special English-like instructions used in a program to tell the computer what to do.

BEAMWIDTH – a measure of the area of sky which a receiving antenna "sees". Small beamwidths are better because the larger the area, the greater the amount of sky noise picked up (see Fig.4.8).

BINARY – of two. A numbering system which has two symbols only, generally designed by 0 and 1 [cf denary (decimal) which has 10 symbols (0, 1, . . . , 8, 9)].

BINARY CODE – a statement in binary digits. In computers a binary code is used to represent letters, numbers and instructions.

BIRD – a colloquial name for an artificial satellite.

BIT ERROR RATE – used in digital systems, it is the ratio of the number of bits received in error to the total number transmitted (Sect.5.5.1).

BORESIGHT – the centre of a transmitting antenna beam.

BRIGHTNESS – see Luminance.

BROADCASTING SERVICE – a radiocommunication service for either sound or television for direct reception by the general public.

BUTTONHOOK FEED – it is possible to mount an LNB on a single rod emanating from the centre of a dish. Because the feedhorn looks back towards the centre of the dish, the guide has a buttonhook shape to ensure that the LNB is situated at the focal point.

CARRIER – a single frequency radio wave which has impressed upon it a band of modulating frequencies (the baseband – see Sect.5.4).

CARRIER-TO-NOISE RATIO – this compares a radio carrier level with the noise level accompanying it. For satellite television C/N should be at least 14 dB. For other services, e.g. military, considerably lower ratios are tolerated.

CASSEGRAIN ANTENNA – an advanced form of parabolic antenna which has a hyperbolic reflector fitted at the focus of the parabola [Fig.4.2(v)].

CATHODE-RAY TUBE (C.R.T.) – an evacuated glass vessel in which an electron beam produces a luminous image on a fluorescent screen, an essential part of television sets and computers.

C.A.T.V. – Community Antenna Television. Signals are received at a cable terminal and fed to subscribers over a cable network (see Fig.7.1).

C-BAND – technically the 4 – 6 GHz microwave band, however in practice running from 3.7 to 6.425 GHz. This band is used mainly in the USA. Note that the actual range quoted by various authorities varies.

C.C.I.R. – French. In English this becomes "International Radio Consultative Committee". It is the international body which sets technical standards for radio transmissions. Operates under the auspices of the International Telecommunications Union.

C.C.I.T.T. – French. In English this becomes "International Telegraph and Telephone Committee". It is the international body which sets technical standards for telegraph and telephone systems. Operates under the auspices of the International Telecommunications Union.

CEEFAX – the teletext service of the BBC.

C.E.S. – Coast Earth Station in the INMARSAT network; connected to the national telephone network and in contact with ships at sea via a satellite.

CHANNEL – the path over which information (data, audio, television) is carried. A channel can be set up on one or more of the following links: an air-path, pair of electrical conductors, terrestrial or satellite radio path, optical fibre.

CHARACTERISTIC IMPEDANCE – of a transmission line. It is the impedance measured at the end of an infinitely long line or a shorter line terminated at the distant end by the characteristic impedance.

CHARGE – is defined as a quantity of electrical energy. It is an invisible certain something possessed only by atomic particles. Within the atom protons are said to have a positive charge, electrons negative. The golden rule is "like charges repel, unlike attract".

CHROMINANCE – the part of a television waveform containing the colour information (Sect.7.1.3).

CIRCULAR POLARIZATION – a type of polarization of a radio wave in which the electric and magnetic fields rotate as the wave travels (Sect.4.5).

CLARKE BELT – the geostationary orbit, named after the writer Arthur C. Clarke who first published the idea of satellites being stationary with respect to earth (Sect.3.2).

C/N – see Carrier-to-Noise Ratio.

COAXIAL CABLE – a special type of cable for use at the higher frequencies (see Fig.1.3(i)).

COMPANDING – a system which uses a volume compressor at the sending end of an audio circuit combined with a volume expander at the receiving end to reduce the effects of noise on a channel (Sect.6.3).

CONIC SECTION – the surface seen when a cone is sliced. Depending on the angle of cut, the section can have circular, elliptical or parabolic shape.

COULOMB – a unit named after Charles Augustin de Coulomb (a French physicist). It is a unit of electric charge and is the quantity of electricity conveyed in one second by a current of one ampere.

COVERAGE AREA – the area of a footprint in which a television picture may be satisfactorily received although freedom from interference from other overlapping transmissions is not guaranteed.

CROSS-COLOUR – a television picture defect which results in swirling coloured patterns (Sect.7.1.5).

CROSS-LUMINANCE – a television picture defect which results in brightness variations at colour changes (Sect.7.1.5).

CRYPTOLOGY – the science or theory of hidden information.

CRYSTAL – (correctly termed "piezo-electric crystal"). A tiny piece of (usually) quartz which is capable of controlling the frequency of an oscillator very accurately (e.g. quartz crystal clocks and watches).

CURRENT (electric) – the passage through a material of electrons. It is measured in *amperes* and for one ampere in only one second, 6.25×10^{18} electrons pass by. One-tenth of one ampere flows through a hand-torch bulb, thousands of amperes are needed for an electric train.

dBW – a unit using decibel notation. It represents a signal level compared to a one watt reference.

DECIBEL – one-tenth of one *bel*. It is a unit used for comparison of power levels in electrical communication (see Appendix 2).

DECLINATION – the angle at which a receiving antenna is set relative to the horizontal, used generally for the alignment of polar mounts (see Fig.7.9).

DECODING – the recovery of the original signal from a coded form of it.

DE-EMPHASIS – part of a technique used especially with frequency modulation systems to reduce the effects of noise on a channel. The improvement arises from pre-emphasis of the higher frequencies at the transmitter with corresponding de-emphasis at the receiver.

DEMAND ASSIGNMENT – a communication system in which channels are assigned to users on demand.

DEMODULATION – the process in which a signal (the baseband) is regained from a modulated carrier wave.

DENARY – of ten. The numerical system we use in daily life, more commonly known as *decimal.*

DEPOLARIZATION – the twisting of the polarization of a radio wave as it travels through the atmosphere.

DEPOLARIZER – a device for the conversion of circularly polarized signals to linear (Sect.4.5).

DESCRAMBLER – an electronic device which restores a scrambled signal to normal (Sect.5.8).

DESPINNING – the arrangement of a spinning satellite to maintain certain antennas in a fixed pointing position.

DEVIATION – a term used in frequency modulation. It is the maximum value of the frequency swing for a particular system.

DIGITAL – a method of handling information by measuring the amplitude of a quantity and coding that amplitude in the binary system.

DIPOLE – short for dipole antenna. A resonant rod antenna of length approximately equal to half a wavelength.

DIRECT BROADCASTING by SATELLITE (DBS) – a radiocommunication service mainly for television in which the signal is directed upwards to a geostationary satellite which then re-transmits it for direct reception by the general public.

DIRECT CURRENT – an electric current which flows continuously in one direction only, for example as provided by a torch or car battery.

DISH – a commonly used name for a parabolic antenna.

DISH DECLINATION OFFSET ANGLE – the angle which is added to the latitude to obtain the declination to which a polar mount is set.

DISH ILLUMINATION – the area of a parabolic antenna as "seen" by the feedhorn. The bounary of the area should coincide with the perimeter of the dish. If the area is greater, i.e. the feedhorn "sees" around the dish, noise may be picked up, if smaller then signal pick-up is reduced.

DISH OFFSET ANGLE – the change in pointing angle for an offset dish compared with a prime focus type (see also Fig.4.2).

DISTORTION – is the change in form or character (other than in magnitude) of a signal during transmission.

D-MAC/D2-MAC – special types of transmission using the MAC format (Sect.7.1.5).

DOWNCONVERTER – a circuit used to translate a modulated signal onto a lower band of frequencies.

DOWNLINK – the radio channel from a satellite to earth.

EARTH STATION – a ground based transmitting and/or receiving installation working to satellites.

E.B.U. – the European Broadcasting Union, an organization comprising the European national broadcasting authorities.

ECCENTRICITY – of a parabola – the degree to which it deviates from circular.

ECHO SUPPRESSOR – a device connected in a 4-wire telephony circuit to suppress echo signals. On detection of a signal in one direction, attenuation is inserted in the opposite direction. Special override facilities are also provided for the occasions when both users are talking together.

ECLIPSE – generally the interception of light from the sun or moon. In satellite working, when the satellite passes through the shadow of the earth and therefore is in darkness (see Fig.3.2).

E.C.S. – European Communications Satellites – these are satellites controlled by EUTELSAT.

EFFECTIVE AREA – of a parabolic antenna is less than its physical area because of various losses. It is the product of the physical area and the efficiency.

EFFECTIVE ISOTROPIC RADIATED POWER (EIRP) – the basis of a technique by which transmitted signal strength can be rated. An isotropic antenna is an elementary theoretical one considered to radiate from a point source equally in all directions (see Fig.4.5).

ELECTROMAGNET – consists of a coil of insulated wire wound round a magnetizable core. The latter becomes magnetized when an electric current flows in the coil but loses its magnetism when the current ceases. The magnetic field is used mainly to operate switches, also to rotate a polarization membrane within a parabolic dish feedhorn.

ELECTROMAGNETIC WAVE – the technical name for a radio wave, so called because it consists of electric and magnetic fields moving in unison (see Sect.1.4).

ELECTRON – a stable elementary particle which carries (or is) a charge of negative electricity. Electrons exist in all atoms and when freed may act as the carriers of electricity. The charge of an electron is insignificant but incredibly large numbers may move together and their charges are additive. It is from the activities of these tiny particles that most of the wonders of electronics arise.

ELECTRON-BEAM – a narrow, pencil-like stream of electrons, all moving at very high velocity in one direction. The beam usually terminates on the screen of a cathode-ray tube and produces a spot of light.

ELECTRON GUN – an electronic device which generates free electrons and concentrates them into a beam. Used mainly in cathode-ray tubes.

ELECTRON-HOLE PAIR – arises when an electron is excited into the conduction band. Two charge-carriers are produced, the free negative electron and the positive "hole" it leaves behind.

ELEVATION – the angle between the horizontal and a line joining an observer to a satellite.

ELLIPSE – has a regular oval shape. It is traced by a point moving in a plane so that the sum of its distances from two other points is constant [see Fig.3.1(ii)].

ELLIPTICAL ORBIT – an orbit having a regular oval shape.

E.L.V. – Expendable Launch Vehicle. A rocket launcher which is used up or destroyed in flight and is therefore not available for re-use.

ENCODING – expressing a message in code, e.g. when characters are stored in binary code in a computer. Often used to describe a process in which the form of an electronic signal is changed.

ENCRYPTION – is a complex secrecy process in which frequencies are jumbled so that the transmission is unintelligble except to the authorized user. It is similar to scrambling except that additional keys are required for decryption.

ENERGY – is the capacity for doing work.

EQUATORIAL ORBIT – an orbit for which the plane includes the equator.

EQUINOX – when the sun crosses the equator and day and night are equal.

EQUIVALENT NOISE TEMPERATURE – in any system all types of noise can be assessed together and quoted by a single value. See also Noise Temperature.

ERROR – usually used with regard to digital transmission. An error occurs when a digital 0 is incorrectly received as a digital 1 and vice versa.

EUTELSAT – European Telecommunications Satellite Organization (Paris). There are 26 member countries.

EVENT TIMER – switches on a satellite receiver and selects channels at times programmed in by the user.

FARAD – is the unit of capacitance (symbol F). It is an inconveniently large unit in practice so we usually talk in terms of microfarads (10^{-6} F), nanofarads (10^{-9} F) and picofarads (10^{-12} F).

FEEDHORN – the device which collects the signal focused onto it by a parabolic antenna. It is in effect a type of waveguide.

FIELD – the sphere of influence of an electric, magnetic or gravitational force. it is undetectable by normal human senses yet is capable of creating a force and creating action. The most commonly encountered field is that of gravity (Sect.1.4.1). It is also a term used in television, i.e. the build-up of a single picture by scanning (Sect.7.1.1).

FLICKER – the rapid increase and decrease of brightness of a television picture.

FLUX – that which flows. The total electric or magnetic field passing through a surface.

FLYBACK – the rapid movement of the spot on a television screen from the end of one line or frame to the beginning of the next (Sect.7.1.1).

FOCUS – the point at which rays meet after reflection or refraction.

FOOTPRINT – the area on land "illuminated" by a satellite radio beam (see Fig.4.7).

FORCE – is a measurable influence tending to cause motion of a body.

FRAME – a single complete television image.

FREQUENCY – the number of cycles per unit time of an electric or electromagnetic signal. The unit is the hertz (Hz) which represents one cycle per second (Sect.1.4.2).

FREQUENCY-CHANGER – an electronic circuit which accepts a modulated wave at one (carrier) frequency and changes it to another frequency but with no change to the modulation.

FREQUENCY DIVISION MULTIPLEX – the combining of two or more input signals at different frequencies into a composite output signal. In practice a multiplex system in which each channel is assigned a different frequency band.

FREQUENCY MODULATION (F.M.) – a method of impressing a signal onto a carrier wave by varying the wave frequency (Sect.5.4).

FREQUENCY SHIFT KEYING (F.S.K.) – a method of modulation used with digital signals. A carrier is switched rapidly between two frequencies to represent a 1 or a 0 (Sect.5.5).

GAIN – a measure of the increase in signal strength when it passes through a system. It can be expressed as the ratio of the power output of the system relative to the power input or more usually as this ratio in decibels (see Appendix 2).

GALLIUM ARSENIDE – is a grey, brittle material used in high-speed semiconductor applications. It has high resistivity and high electron mobility.

GEOSTATIONARY – stationary with respect to earth. In a geostationary orbit satellites complete one revolution in the same time as the earth does. Each satellite therefore moves so that it always remains above the same point on the earth's surface (see Fig.3.1).

GEOSYNCHRONOUS – an earth orbit for which the time required for a satellite to complete one revolution is an integral fraction of a sidereal day.

GIGAHERTZ (GHz) – a unit of frequency equal to one thousand million hertz (10^9 Hz).

GLOBAL POSITIONING SYSTEM (G.P.S.) – a system of satellites controlled by INMARSAT for general navigation.

GRAVITY – the attractive force between any two bodies having mass. There is no explanation of this force, it has been provided by Nature in order to keep all things together.

GREGORIAN ANTENNA – an advanced form of parabolic antenna. It employs two reflecting surfaces and is usually offset. Because there is no blocking of the radio wave by a sub-reflector, high efficiencies are obtained [see Fig.4.2(vi)].

G/T (GAIN/TEMPERATURE) – is a figure of merit describing the ratio of antenna gain to the system noise temperature. The unit is decibels per degree Kelvin.

GYROSCOPE – a device used for keeping satellites, ships and aircraft in an equilibrium position.

HALF POWER BEAMWIDTH (H.P.B.W.) – the beamwidth angle of a transmitting antenna which produces a footprint contour on which the signal power is 3 dB lower than the maximum value (see Fig.4.8).

HARMONIC – a component of a wave having a frequency which is an integral number of times that of the basic (fundamental) frequency, e.g. if the fundamental frequency is denoted by f Hz, then its harmonics are $2f$, $3f$, $4f$, etc.

HDTV – High Definition Television – newer systems being developed having a greater number of lines than the current 625 to provide a sharper picture.

HEAVENLY ARC – the path of the sun as seen from earth during one day.

HERTZ – the international standard unit of frequency equal to one cycle per second (after Heinrich Hertz, a German physicist).

HOLE – when an electron is displaced from an atom, the latter effectively becomes a positive charge and its "vacancy" for an electron is known as a hole.

HOP – the distance travelled through the atmosphere or space between two positions on earth.

HORIZONTAL POLARIZATION – a radio wave for which the electric field is horizontal and the magnetic field vertical (Sect.1.4).

HORIZON-TO-HORIZON – a special polar mount with a greater than average range of movement.

HORN (electromagnetic) – a horn-shaped termination on a waveguide used in both transmitting and receiving paraboloid antennas. Generally the orifice is of rectangular or circular shape (Sect. 4.2.2).

H.P.A. (HIGH POWER AMPLIFIER) – a linear output power amplifier used at microwave frequencies.

IMPEDANCE – expresses the degree of opposition a circuit presents to the passage of an alternating electric current. It is measured in *ohms.*

INCLINATION – the angle between the orbital plane of a satellite and the equator.

INCLINOMETER – an instrument used to measure angles of elevation.

INMARSAT – the international organization for communication with maritime, aeronautical and land-based mobiles.

INTELSAT – the International Telecommunications Satellite Organization, a body controlling the international satellite system for telephony, data and television.

INTERMEDIATE FREQUENCY (I.F.) – a frequency to which that of a modulated carrier wave is reduced for processing.

ION – the particle which remains when a neutral atom gains or loses one or more electrons, i.e. an atom without its normal complement of electrons.

IONIZATION – the process which produces a positive or negative charge on an atom originally electrically neutral by the removal or addition of one or more electrons.

IONOSPHERE – the ionized region of the upper atmosphere capable of returning radio waves of a range of frequencies (lower than those used for satellite working) back to earth (Sect.1.1).

I.R.D. – Integrated Receiver/Decoder. A satellite receiver with a decoder built in.

ISOTROPIC ANTENNA – a reference antenna radiating energy from a point source in all directions (see Fig.4.5).

I.T.U. – International Telecommunications Union – concerned with international standards for radio, telegraphy and telephony. It is the body responsible for allocating frequencies to be used for satellite working world-wide.

KELVIN – a degree of temperature equal to a Centigrade or Celsius degree. The Kelvin scale however starts at absolute zero so 0° Celsius is equivalent to 273° on the Kelvin scale (after Lord Kelvin, a British physicist).

KEYING – the making and breaking of an electrical circuit for the transmission of information. This was originally by a key operated by hand (e.g. for Morse Code), nowadays usually electronically.

KINETIC ENERGY – the energy possessed by a body due to its weight and motion.

Ku-BAND – the 11 – 14 GHz microwave band (Europe). There are variations of this, e.g. 10.7 – 18.0 GHz and others.

L-BAND – the frequency range 1 – 2 GHz (USA).

L.E.D. – Light-Emitting Diode. Usually a very small electric lamp having no filament. The glow is normally but not necessarily red. These small lamps are used mainly as indicators or as the basic units of an illuminated letter or number display.

L.H.C.P. – left-hand circular polarization – of a radio wave. Looking towards an oncoming wave, the rotation is anti-clockwise (Sect. 4.5).

LINK MARGIN – on a satellite downlink, the degree by which the carrier power exceeds the minimum level for a satisfactory television picture.

L.N.A. – low noise amplifier.

L.N.B. – low noise block converter. This is an electronic device at the focus of a parabolic antenna which amplifies the incoming radio wave and converts it to a lower intermediate frequency for transmission over a cable to the satellite receiver (Sect.7.2.1).

L.N.C. – low noise converter – see L.N.B.

LOOK ANGLE – the azimuth or elevation angle used to point an antenna towards a satellite. It varies both with the position of the observer and of the satellite.

LOSS – a measure of the extent to which the amplitude of a signal is decreased by its passage through a system, usually expressed in decibels.

LUMINANCE – the luminous intensity or amount of white light emitted from a small area on a television screen. With regard to a television waveform, it is that part which contains the brightness information (Sect.7.1.3).

MAC – Multiplexed Analogue Components, a partly analogue, partly digital system of television transmission. It produces pictures of enhanced quality compared with the PAL and SECAM systems.

MASS – the quantity of matter (that which occupies space) in a body as measured by its acceleration under a given force.

MAGNETIC NORTH – the earth can be considered as being a huge magnet having North and South magnetic poles. The line joining these poles is inclined slightly to the axis of rotation hence true North and magnetic North do not coincide. A compass needle points to magnetic North.

MAGNETIC DEVIATION – see Magnetic Variation.

MAGNETIC POLARIZER – an electronic device controlled by a current from the receiver for changing polarity in an LNB.

MAGNETIC VARIATION – the angular difference at any place between true North and magnetic North.

MANPACK – a portable miniature satellite terminal used by the army.

MASS – the quantity of matter a body contains as measured by its acceleration when a given force is applied.

MEMORY – an electronic store of information. For example a satellite receiver can store in its memory data with regard to television channels such as frequency, azimuth, elevation, polarization. The information is stored in binary.

MERIDIAN (true meridian) – a circle on the earth's surface of constant longitude passing through a given place and the North and South poles.

MICROWAVE – an ultra-short wave of wavelength less than about 30 cm (frequency = 1 GHz).

MIXER – an electronic device which accepts two different frequencies at its input and produces a combination of these frequencies at the output.

MODULATION – the process in which a signal (the baseband) is impressed upon a higher frequency carrier wave (Sect.5.4).

MOONBOUNCE – radio transmission achieved by using the moon as a reflector for returning signals to earth (Sect.1.1.1).

MULTI-FEED – an antenna system with two (or more) LNB's fitted to one fixed dish so that more than one satellite can be viewed.

MULTIPLE ACCESS – a system which provides several users with access to the same channel.

MULTIPLEX – transmission of two or more separate elements over a single channel.

N.A.S.A. – National Aeronautics and Space Administration. The USA organization responsible for the space exploration programme.

NEGATIVE – the name given to the electrical charge of the electron.

NOISE – any unwanted electrical or audio signal which accompanies but has no relevance to the transmitted signal (Sect.1.5).

NOISE-FACTOR – is defined as the relationship between the input and output signal-to-noise power ratios of any system. It therefore indicates the extent to which the noise generated within an equipment degrades the signal-to-noise ratio as the signal passes through.

NOISE FIGURE – the noise factor in decibels. For example the noise figure of an LNB shows its contribution of unwanted noise, the lower the noise figure, the better.

NOISE TEMPERATURE – a method of assessing electrical noise. It is the temperature in degrees Kelvin to which the noise source would have to be raised to produce the same noise output as the system itself.

N.T.S.C. – National Television Standards Committee. This also refers to the 525 line/60 field system used for television broadcasting in the USA and some other countries.

NUTATION DAMPING – the process of damping wobbling motion occurring in the spin axis of a satellite.

OFFSET ANGLE – the angle which is added to the latitude of a receiving installation for setting the declination angle to line up a polar mount with the geostationary arc (Sect.7.3.1.4).

OFFSET ANTENNA – a special design of parabolic antenna in which the LNB is outside the path of the incoming signal. Generally the feedhorn is below the centre of the dish (see Fig.4.2).

ON-SCREEN DISPLAY – in satellite television. Channel and tuning data are presented on screen over the picture.

OPTICAL FIBRE – a very fine circular strand of glass, about human hair thickness. Such a fibre transmits electromagnetic waves at frequencies of light with very high beamwidths. Optical fibres are used for the transmission of telephony, data and television.

ORACLE – a teletext magazine broadcast by the UK Independent Broadcasting Authority.

ORBIT – the closed path (usually circular or elliptical) followed by a planet, spacecraft or satellite around a larger body.

OSCILLATOR – an electronic device for the production of alternating electric currents, i.e. waveforms having frequencies from one or two up to many thousands of millions of reversals per second.

PACKET – a sequence or block of binary digits.

PAL – Phase Alternation Line – the 625 line/50 field television system used by many European countries (not France).

PANDA – a system of audio processing which improves the signal-to-noise performance of narrow-band audio sub-carriers. It uses a combination of companding with pre- and de-emphasis.

PARABOLA – a curve obtained by slicing a cone at a certain angle. It conforms to the mathematical equation $y^2 = 4fx$ where f is the distance of the focus from the centre (focal length) – see Fig.4.1.

PARABOLIC ANTENNA – a dish type of antenna having a curved surface of parabolic shape. A receiving type has the capability of reflecting an incoming radio wave and focusing its energy onto a single point (Sect.4.2.1).

PARENTAL LOCK – a special electronic locking arrangement fitted to a satellite receiver to deny access to one or more specified channels.

PARITY CHECKING – a system used in digital transmission for the detection of errors (Sect.5.5.1).

PASSIVE SATELLITE – a satellite capable of reflecting electromagnetic waves only.

PAYLOAD – the productive or useful part of the load of a rocket.

PERIGEE – the point nearest to the earth in the orbit of, for example, the moon, a planet or a satellite. A geostationary orbit is circular and therefore has no perigee.

PERMEABILITY – is a measure of the ease with which a material can be magnetised.

PERMITTIVITY – is a measure of the ability of a material to store electrical energy when it is situated in an electric field.

PERSISTENCE OF VISION – an image in the eye persists for a period of time (up to about 0.1 seconds) – Sect.7.1.3.

PETALIZED – a form of construction of dish antennas by using identical metal "petals" bolted together to form the dish instead of the more usual solid construction.

PHASED ARRAY – a number of antennas spaced apart and interconnected in such a way that their outputs are all in phase.

PHOTON – is defined as a *quantum* of energy, i.e. a discrete, infinitesimally small amount. We use the idea to explain electromagnetic phenomena when applied to light.

PHOTOVOLTAIC – the production of electricity from two substances in contact and exposed to light.

PICTURE ELEMENT – the smallest area of a television picture which can be displayed.

PIXEL – see Picture Element.

POLAR CURVE – one which is related in a particular way to a given curve and to a fixed point called a pole. The sun follows a polar curve each day.

POLARIZATION – the way in which the electric field of a radio wave is disposed relative to the direction of propagation (Sect.4.5).

POLARIZATION OFFSET – is required when the polarization planes of satellite and receiving antenna are not in line. The amount of rotation of an LNB required to line them up is the polarization offset.

POLARIZER – a device in the LNB waveguide which converts between the various incoming signal polarizations. Generally there are three types: (i) electromechanical, operated by a motor; (ii) magnetic, the incoming wave is twisted by a magnetic field; (iii) switched by a solid-state device. All are operated from the indoor receiver.

POLAR MOUNT – a special type of parabolic antenna mounting which allows the antenna to rotate and at the same time adjust its elevation so as to follow the geostationary arc (Sect.4.8.1.2).

POLAR ORBIT – the orbit of a satellite when its plane includes both North and South poles.

POLAROTOR – an electronic device which rotates an LNB for correct alignment with the polarization of the incoming radio wave. Alternatively a membrane built into the feedhorn may be moved into the required positions by an electromagnet.

POLE – is best described as a fixed reference point. We are mainly concerned with the North and South poles of the earth and of magnets, also the positive and negative poles of an electricity supply or battery. A pole is also the highest point in a geostationary arc.

POSITIONER – a device which controls the motor of a motorized dish so as to point the dish in a predetermined position.

POSITIVE – the name given to an electrical charge opposite to that of the electron.

PRE-AMPLIFIER – an amplifier which raises a low-level signal to a value suitable for driving a main amplifier.

PRE-EMPHASIS – part of a technique used especially with frequency modulation systems to reduce the effects of noise on a channel. The improvement arises from pre-emphasis of the higher frequencies at the transmitter with corresponding de-emphasis at the receiver.

PRIME FOCUS ANTENNA – a parabolic antenna with the feedhorn mounted at the focus (see Fig.4.2).

PROPAGATION PLANE – of an electromagnetic wave. It is that plane which contains both the direction of the wave and the direction of the electric field.

PROPAGATION TIME – the time taken by a signal (usually an electromagnetic wave) to travel between two fixed points.

PROPELLANT – fuel used in a rocket engine to provide thrust (Sect.2.1).

PROTON – the atom is a complex arrangement of positive, negative and neutral particles bound together. Protons are particles contained within the nucleus or core of the atom, each having the same charge as an electron but positive. An electrically neutral atom has the same number of protons as electrons, hence the charges balance.

QUADRATURE PHASE-SHIFT KEYING (Q.P.S.K.) – a type of digital signal modulation (Sect.5.5).

QUANTIZATION NOISE – noise generated within a pulse-code system due to the fact that the exact level of a waveform at any instant cannot be determined. The noise is reduced as the number of sampling levels is increased.

QUANTIZING – the process of representing the instantaneous amplitudes of a waveform by steps.

RADIAN – the angle at the centre of a circle subtended by an arc equal in length to the radius. It is equivalent to 57.3 degrees.

RADIATION – the outflow of energy in the form of a radio wave, generally from a radio antenna.

RADIOMETER – for the measurement of radiant energy.

RADOME – a dome or other covering for the protection of satellite and radar equipment.

RANDOM NOISE – electrical noise generated by the continual movements of a large number of free electrons in a conductor.

RASTER – a pattern of scanning lines on a television screen (Sect.7.1.1).

REGENERATIVE REPEATER – a repeater which accepts a mutilated digital signal and transmits onwards a "perfect" replica of it.

REPEATER – an electronic device for automatic re-transmission or amplification of a signal.

RESIDUAL CURRENT CIRCUIT BREAKER – a device connected in the domestic mains supply which disconnects the supply quickly on sensing an earth fault.

RESISTANCE – expresses the degree of opposition a circuit presents to the passage of a direct current. It is measured in ohms.

R.H.C.P. – right-hand circular polarization of a radio wave. Looking towards an oncoming wave, the rotation is clockwise (Sect. 1.4.1).

ROCKET – an engine operated by the reaction due to a continuous jet of expanding gases.

SAMPLING – measures the amplitude of an analogue signal at regular intervals for conversion to digital form.

SATCOM – a series of satellites providing services in the USA.

SATELLITE – a heavenly or artificial body revolving around another larger one.

SATELLITE NEWS GATHERING (S.N.G.) – a news system employing portable equipment easily set up for transmitting over an uplink.

S-BAND – the frequency range 2 – 4 GHz (USA).

SCANNING – the resolution in a prearranged pattern of a television picture into its elements of light, shade and colour (Sect.7.1.1).

SCART – the European 21-pin plug and socket in general use for linking audio and visual equipment.

SCRAMBLING – is the method by which frequencies in a transmission are jumbled so as to make the outcome unintelligble except to the intended recipient (Sect.5.8). See also Encryption.

SECAM – Système En Couleurs à Mémoire, the 625 line/50 field television system used in France, Luxembourg and Monaco.

SENSITIVITY – is generally the response given to a measurable stimulus. For a satellite receiver it is the smallest received signal capable of producing a certain specified output.

SERVICE AREA – the area within a footprint in which the signal is sufficiently strong for a satisfactory and interference-free television picture to be received.

S.E.S. – Ship Earth Station – a ship in contact with land via a satellite in the INMARSAT network.

S.E.S. – Société Européenne des Satellites – the consortium based in Luxembourg which is responsible for the ASTRA satellites.

SHUTTLE – the re-usable space vehicle manned by astronauts which is capable of carrying satellites up into space for launching into orbit. The choice for N.A.S.A.'s space programme.

SIDEREAL DAY – the period of the earth's rotation with respect to the stars (about 4 minutes shorter than a solar day – Sect.3.2).

SIGNAL – an intelligible sign conveying information. In satellite television signals are in the form of electronic waveforms varying with time. There are radio, video and audio signals.

SIGNAL-LEVEL METER – a measuring device, usually based on a needle moving over a scale, which can be tuned to a signal. The travel of the needle across the scale indicates the strength of the signal being measured.

SIGNAL-TO-NOISE RATIO (S.N.R. or S/N) – a method of indicating the strength of a signal compared with that of the noise accompanying it, usually expressed in decibels.

SKEW – the fine tuning of polarity to account for the differing angles of arrival of signals from the various satellites.

SKYNET – a range of satellites engaged in defence communications.

SMART CARD – a small card with electronic data printed on it in a magnetic stripe which is inserted into a decoder for descrambling a protected channel.

SMATV – Satellite to Master Antenna Television. Signals are received on a single antenna and fed by cable to apartments, throughout a hotel, group of houses, etc. (see Fig.7.1).

S.N.G. – Satellite News Gathering. The transmission of news from small portable uplink stations to the central news collection point (Sect.8.1.4).

SOLAR – a period of time determined by the sun, e.g. solar day, solar year.

SOLAR CELL – a cell which converts radiation from the sun into electricity.

SPARKLIES – a colloquial term for white dots or flashes on a television screen, usually the result of a weak signal.

SPEECH INTERPOLATION SYSTEM – a telephony system in which a user has access to a channel only while actually talking.

SPINNER – a satellite which spins to stabilize its attitude.

SPOT BEAM – a narrow satellite radio beam. The footprint is therefore smaller than for a normal beam (Sect.4.6).

STEREOPHONIC REPRODUCTION (Stereo = three dimensional) – an audio reproduction system using two or more loudspeakers spaced apart to provide the illusion of sound arriving from all directions. It relies on the fact that the human brain is sensitive to sound direction and the system therefore provides a more realistic sound signal.

S.T.S. – Space Transport System, the American space "Shuttle", a re-usable launch vehicle.

SUB-CARRIER – a carrier associated with a satellite television signal carrying the audio and possibly other radio transmissions.

TELECONFERENCE – a conference which uses a satellite system to relay video and audio signals between locations anywhere on earth.

TELEMETRY – taking readings of a measuring instrument at a distance, usually via a radio link.

TELEPORT – a small earth terminal working directly to a satellite and serving a local business community.

TELETEXT – separate information transmitted with a television picture signal and which can be displayed on the screen in place of the normal picture. In the UK the BBC calls it CEEFAX, the IBA uses the name ORACLE.

TELETYPE – short for "teletypewriter" – typewriting from a distance.

TERMINATING SET – used in telephony. A circuit consisting mainly of transformers used to couple a 2-wire telephone line to a 4-wire system (see Fig.6.2).

THERMAL NOISE – electrical noise which arises from the agitation of electrons in a conductor due to heat (Sect.1.5.1).

THRESHOLD – when applied to a satellite receiver, this is a measure of its sensitivity and generally refers to the carrier-to-noise threshold which is used to determine picture quality with weaker signals. It is the lowest signal level from the antenna which the receiver can convert into a satisfactory picture (hence the lower the better).

TIME BASE – electronic equipment for generating a repetitive timing voltage.

TIME DIVISION MULTIPLEX (T.D.M.) – a system in which two or more input signals are transmitted separated by time. The full bandwidth of the system is made available at regular intervals to each of the separate input signals.

TERRESTRIAL – of the earth (Latin, *terra* = earth).

TONNE – the metric measurement of weight equal to 1000 kilograms (roughly the same as the Imperial ton).

TRACKING – plotting the path of a moving object.

TRANSCODER – an electronic circuit for translating from one television format to another, e.g. PAL to SECAM.

TRANSFER ORBIT – an orbit in which a satellite is first placed prior to being moved into another (usually the final) orbit (Sect. 2.1.1).

TRANSLATION – the shifting of an amplitude or frequency modulated signal up or down in frequency.

TRANSMITTER – the equipment used to transmit speech, data, television by radio wave via an antenna.

TRANSPONDER – an electronic system in a satellite which receives a signal from earth, changes its frequency, then amplifies it for transmission back to earth.

TRAVELLING-WAVE TUBE AMPLIFIER (T.W.T.A.) – a microwave amplifier in which very high frequency waves travel along a tube and as they do so are increased in amplitude. Used in satellites (Sect.5.7).

TRIGGER – a short duration electrical pulse which initiates some action.

T.T.C. – Telemetry, Tracking and Command. These are the facilities required by a ground control station for monitoring the spatial and electrical conditions of a satellite and for signalling back instructions.

TUNED CIRCUIT – an electronic circuit which resonates or "tunes" to one particular frequency (or narrow band of frequencies) only.

T.V.R.O. (TELEVISION RECEIVE-ONLY) – a single (usually domestic) satellite television installation (Fig.7.1).

UPCONVERTER – a circuit used to translate a modulated signal onto a higher band of frequencies.

UPLINK – the radio channel from earth up to a satellite.

VERTICAL POLARIZATION – a radio wave in which the electric field is vertical and the magnetic field horizontal (Sect.4.5).

VIDEO – that which we can see. A term used in the broadcasting of television pictures. A video waveform is the band of frequencies representing the output of a television camera.

VIDEOCRYPT – one of the "smart card" encryption systems.

VIDEOTEX – a system in which signals are transmitted over telephone lines for ultimate display on the screen of a television receiver.

VIEWDATA – see Videotex.

VISION SIGNAL – in television, a carrier wave modulated by the video signal or waveform.

VOLT – a measure of the difference of electric potential in a circuit. Car batteries are 12 V, the electricity mains some 240 V, lightning is many millions.

VSAT – very small aperture terminal. This is a type of relatively inexpensive earth station which makes possible two-way communication from remote locations. It also provides access to private networks involving up to thousands of terminals – rather similar facilities to those provided by a terrestrial telephone network.

W.A.R.C. – World Administrative Radio Conference. A conference was called in 1977 to allocate satellite frequencies to the various nations of the world. Operates under the auspices of the International Telecommunications Union.

WATT – the unit of electric power. Only one watt or so is required to light a hand-torch bulb but several thousand watts are required for an electric cooker.

WAVEFORM – the shape obtained by plotting the amplitude of a varying quantity against time on a graph – see for example Figures 1.4 and 4.6.

WAVEGUIDE – a metal tube of rectangular or circular cross-section through which microwaves are transmitted – see Figure 1.3.

WAVELENGTH – the distance between two similar points in a sound wave or electromagnetic wave.

WEGENER – a noise reduction system used with satellite television.

X-BAND – the frequency range 8 – 12.5 GHz (USA).

Appendix 2
USING DECIBELS

The decibel (dB) is a unit in a logarithmic system of calculations used to express power, current or voltage *ratios*. The advantages of the system are:

(1) large numbers are reduced and so become more manageable;

(2) in a complex system where different gains or losses occur, calculation of the overall power etc. ratio by multiplication is unwieldy. By expressing each ratio in a logarithmic unit, addition takes the place of multiplication, a more manageable process altogether.

Alexander Graham Bell, the Scottish-American inventor first introduced the *bel* which is simply the common logarithm of the power ratio. This gave rather low figures so an offshoot soon appeared, the *decibel*, which is one-tenth of one bel. Hence:

Power:

$$\text{No. of decibels} = 10 \log_{10} \frac{\text{power sent } (P_s)}{\text{power received } (P_r)}$$

Voltage and Current:

$$\text{No. of decibels} = 20 \log_{10} \frac{\text{voltage or current at input}}{\text{voltage or current at output}}$$

(input and output measurements must be in the same impedance).

If no logarithm tables or suitable calculator or computer are available, Table A2.1 can be used to determine the approximate power ratio for any dB value. Some examples are:

(i) 17 dB = 10 dB + 7 dB converted to power ratio =

$$10 \times 5.01 = \underline{50.1}$$

(ii) 35 dB = 30 dB + 5 dB converted to power ratio =

$$1000 \times 3.16 = \underline{3160}$$

(iii) 206 dB = 200 dB + 6 dB converted to power ratio =

$$\underline{10^{20} \times 3.98}$$

this can remain in scientific notation as $\underline{3.98 \times 10^{20}}$

(iv) –59 dB = –60 dB + 1 dB converted to power ratio =

$$10^{-6} \times 1.259 = \underline{1.259 \times 10^{-6}}$$

TABLE A2.1

APPROXIMATE RELATIONSHIP BETWEEN DECIBELS AND POWER RATIOS

dB	Power ratio P_s/P_r	dB	Power ratio P_s/P_r
–10	0.1	5	3.16
– 5	0.32	6	3.98
– 3	0.5	7	5.01
0	1.0	8	6.31
0.2	1.05	9	7.94
0.5	1.12	10	10
1.0	1.26	20	100
2	1.58	30	1000
3	2.0	40	10^4
4	2.51		↓

thereafter for multiples of 10 dB the exponent of 10 in the ratio is the same as the tens figure of the dB value, e.g. 200 dB = 10^{20}, –200 dB = 10^{-20}.

Appendix 3

FREQUENCY BAND CLASSIFICATION

The internationally agreed classification of frequency bands is given in the table below. The *band number* is the exponent of 10 used in converting the range 0.3 – 3 Hz to that required, e.g. band number n covers the range 0.3×10^n to 3×10^n Hz.

Band No.	*Frequency Range*	*Description*	*Letter Designation*	*Metric Subdivision*
4	3 – 30 kHz	Very low frequency	VLF	Myriametric
5	30 – 300 kHz	Low frequency	LF	Kilometric
6	300 kHz – 3 MHz	Medium frequency	MF	Hectometric
7	3 – 30 MHz	High frequency	HF	Decametric
8	30 – 300 MHz	Very high frequency	VHF	Metric
9	300 MHz – 3 GHz	Ultra high frequency	UHF	Decimetric
10	3 – 30 GHz	Super high frequency	SHF	Centimetric
11	30 – 300 GHz	Extremely high frequency	EHF	Millimetric

Appendix 4

CONVERSION OF MINUTES TO A DECIMAL FRACTION OF A DEGREE

Frequently latitudes and longitudes are quoted as so many degrees and minutes. However generally in satellite work degrees with decimal fractions are preferred. Figure A4.1 is a simple nomogram for quick conversion but with sufficient accuracy for most work.

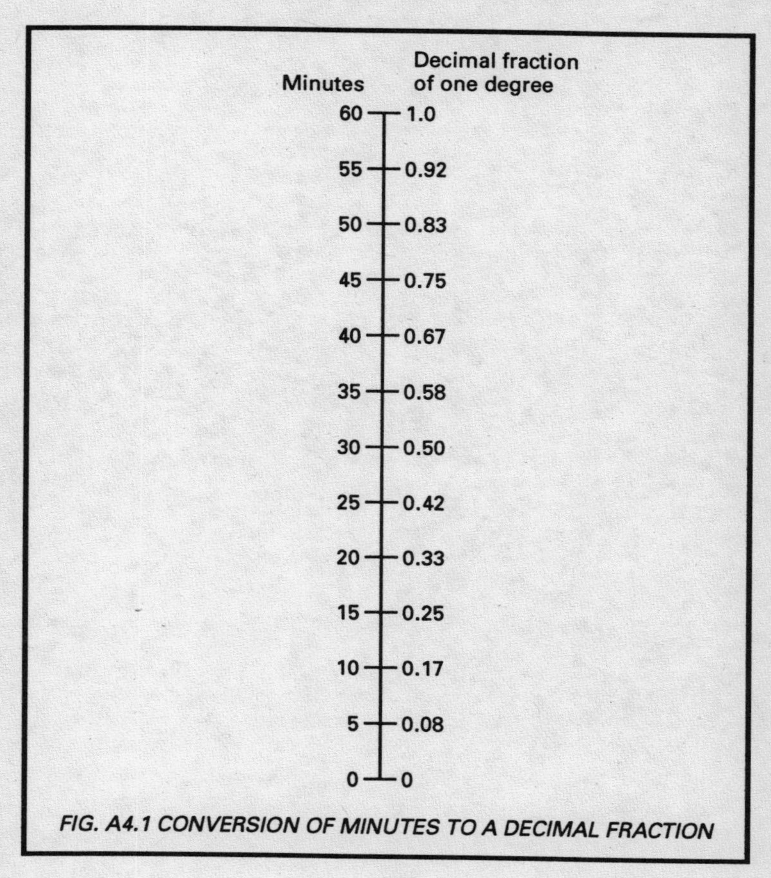

FIG. A4.1 CONVERSION OF MINUTES TO A DECIMAL FRACTION

Appendix 5
TRANSMISSION FORMULAE

This appendix considers some of the transmission aspects of satellite communication in slightly greater detail.

A5.1 Noise

The subject of electrical noise is introduced in Section 1.5.1 and is considered again in Section 5.2. The study and assessment of noise is of paramount importance because noise primarily determines minimum signal levels.

A5.1.1 Figures and Factors

A common way of specifying noise performance is by the *noise figure.* However consider *noise factor* first. This is defined as the relationship between the input and output signal-to-noise power ratios and it indicates how much the noise generated within an equipment degrades the ratio as the signal passes through. Then:

$$\text{Noise factor, } F = \frac{P_{si}/P_{ni}}{P_{so}/P_{no}}$$

where P_s and P_n represent signal and noise powers and i and o refer to input and output.

For example, if the input ratio is 6 (i.e. the signal has 6 times the power of the noise) and at the output the ratio is 4, then:

$$F = \frac{6}{4} = 1.5 \, .$$

If the output ratio is reduced to 3 because greater noise is generated within the equipment, then:

$$F = \frac{6}{3} = 2$$

so the higher the value of the noise factor, the worse the equipment is from the point of view of internal noise generation.

Noise *figure* follows for it is simply noise factor expressed in decibels, i.e.

$$F_{dB} = 10 \log F$$

and the two noise factors above become noise figures of:

$$10 \log 1.5 = 1.76 \text{ dB}$$

and

$$10 \log 2 = 3.01 \text{ dB}.$$

In the next section we see how the internally generated noise increases directly with *absolute* temperature, hence the temperature at which a noise figure is measured should be quoted. Take for example two specifications for LNB's: A has a noise figure of 2.5 dB at 25°C; B's is 2.5 dB at 60°C. To compare them adjustments must be made for temperature so let us check what happens if the temperature surrounding A is raised from 25 to 60°C. The absolute (or Kelvin) and Celsius temperature scales have the same size degrees but the formula works in degrees Kelvin which starts at absolute zero, i.e. 273 degrees below the Celsius:

$$25^\circ\text{C} = 273 + 25 = 298 \text{ K}$$

$$60^\circ\text{C} = 273 + 60 = 333 \text{ K}$$

(the degree symbol is not used with the Kelvin scale). Then increase in noise is equal to:

$$\frac{333}{298} = 1.12,$$

i.e. about half of one decibel. Accordingly the comparison is more accurately stated as:

Noise figure of A at 60° C = 3.0 dB

Noise figure of B at 60° C = 2.5 dB

so B has the better noise performance.

Believe it or not, this is the simplest of noise evaluation methods. Communications engineers have more flexible ways of assessing system noise performance such as rating even galactic noise by an "equivalent noise temperature" as developed below.

A5.1.2 Equivalent Noise Temperature

Of the sources of noise contributing to the total noise at the output of a satellite channel, only that due to thermal agitation can be described by a formula with any pretence to accuracy.

From Boltzmann's basic relationship between energy and temperature, the *available* noise power P_n generated by the thermal agitation of electrons is given by:

$$P_n = kTB \text{ watts}$$

where k is Boltzmann's constant (1.38×10^{-23} J/K – joules or watts per hertz per degree Kelvin); T is the absolute temperature of the noise source in degrees Kelvin; and B is the bandwidth in Hz (for an amplifier usually taken as the upper and lower 3 dB points).

Note the word "available" in P_n. It is the maximum power which can be obtained on the basis of matched conditions between noise source and load.

Satellite communication demands low noise because of the weak signals involved. For such systems the concept of *noise temperature* is found to be most convenient. It can apply to a single item or to a whole system. The Equivalent System Noise Temperature, T_s is the temperature to which a resistance connected to the input of a noise-free receiver would have to be raised in order to produce the same output noise power as the original receiver.

T_s is therefore an overall measure of noise performance and in the case of a satellite receiving installation can account for normal device thermal noise, radio noise picked up by the dish and thermal noise of the dish itself. Consider an LNB connected to a receiving dish as in Figure A5.1. The noise contribution of each unit is shown. P_{no}, the total noise power output is their sum. Then:

$$P_{no} = GkT_sB$$

where G is the total gain. By equating this with the total noise power as shown in Figure A5.1 and rearranging:

$$T_s = \left(T_a + T_r + \frac{T_m}{G_r} + \frac{T_i}{G_rG_m}\right)$$

showing that proceeding along the chain, the later stages have less effect, a not unexpected conclusion because they are followed by lower gain.

There is also a simple relationship between noise figure and effective noise temperature, T_n for any device:

$$T_n = (F-1)T$$

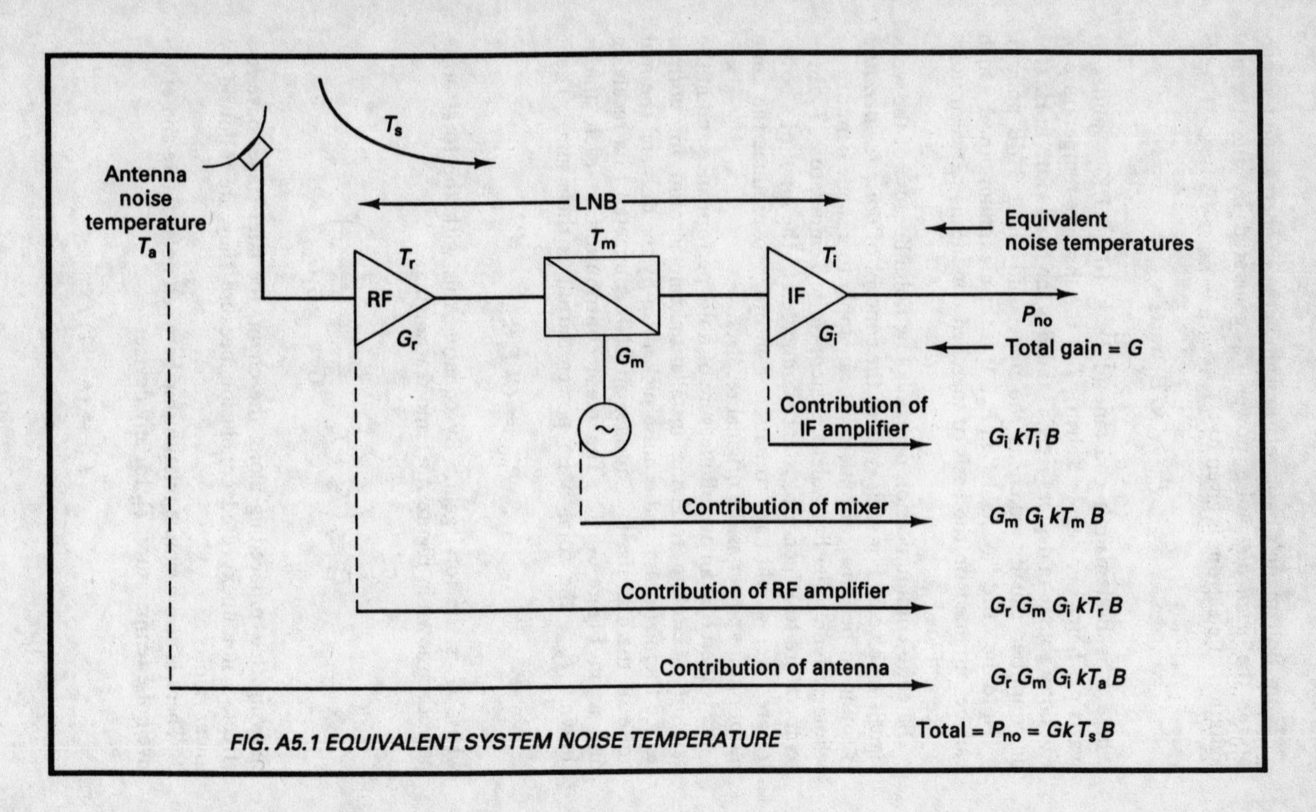

FIG. A5.1 EQUIVALENT SYSTEM NOISE TEMPERATURE

where T is the reference temperature (usually 290 K) and F is the noise figure (at 290 K).

As an example, consider an LNB with a noise figure of 1.8 dB (= 1.51), then

$$T_n = (1.51 - 1) \times 290 = 148 \text{ K}.$$

Appendix 6

HYPOTHETICAL GAIN OF PARABOLIC ANTENNAS (dB)

Especially for readers who do not have access to a computer or scientific calculator, this appendix quotes the calculated gains of parabolic antennas of known efficiencies over the European television range for antenna diameters up to 1.5 metres. The formula used is that quoted in Section 4.4 [Equation (2)] and for convenience is reproduced here, i.e.

$$\text{hypothetical gain, } G_r = 10 \log (0.00011 \times D^2 f^2 \eta) \text{ decibels}$$

where D is the diameter of the parabolic antenna in centimetres, f is the transmission frequency in GHz, and η is the percentage efficiency of the antenna.

A little guesswork or interpolation may be required for values other than those shown.

Table A6.1 – HYPOTHETICAL GAIN OF PARABOLIC ANTENNAS (dB)

Frequency (GHz)	*Wavelength* (cm)	*Diameter* (cm)										
		35	60	70	80	90	100	110	120	130	140	150
EFFICIENCY, η = 50%												
11.0	2.73	29.1	33.8	35.1	36.3	37.3	38.2	39.1	39.8	40.5	41.2	41.8
11.2	2.68	29.3	34.0	35.3	36.4	37.5	38.4	39.2	40.0	40.7	41.3	41.9
11.4	2.63	29.4	34.1	35.4	36.6	37.6	38.5	39.4	40.1	40.8	41.5	42.1
11.6	2.59	29.6	34.3	35.6	36.8	37.8	38.7	39.5	40.3	41.0	41.6	42.2
11.8	2.54	29.7	34.4	35.7	36.9	37.9	38.8	39.7	40.4	41.1	41.8	42.3
12.0	2.50	29.9	34.5	35.9	37.0	38.1	39.0	39.8	40.6	41.3	41.9	42.5
12.2	2.46	30.0	34.7	36.0	37.2	38.2	39.1	39.9	40.7	41.4	42.0	42.6
12.4	2.42	30.2	34.8	36.2	37.3	38.3	39.3	40.1	40.8	41.5	42.2	42.8
12.6	2.38	30.3	35.0	36.3	37.5	38.5	39.4	40.2	41.0	41.7	42.3	42.9
12.8	2.34	30.4	35.1	36.4	37.6	38.6	39.5	40.4	41.1	41.8	42.5	43.1

EFFICIENCY, η = 60%												
11.0	2.73	29.9	34.6	35.9	37.1	38.1	39.0	39.9	40.6	41.3	41.9	42.5
11.2	2.68	30.1	34.7	36.1	37.2	38.3	39.2	40.0	40.8	41.5	42.1	42.7
11.4	2.63	30.2	34.9	36.2	37.4	38.4	39.3	40.2	40.9	41.6	42.3	42.9
11.6	2.59	30.4	35.0	36.4	37.5	38.6	39.5	40.3	41.1	41.8	42.4	43.0
11.8	2.54	30.5	35.2	36.5	37.7	38.7	39.6	40.4	41.2	41.9	42.5	43.1
12.0	2.50	30.7	35.3	36.7	37.8	38.9	39.8	40.6	41.3	42.0	42.7	43.3
12.2	2.46	30.8	35.5	36.8	38.0	39.0	39.9	40.7	41.5	42.2	42.8	43.4
12.4	2.42	30.9	35.6	37.0	38.1	39.1	40.1	40.9	41.6	42.3	43.0	43.6
12.6	2.38	31.1	35.8	37.1	38.3	39.3	40.2	41.0	41.8	42.5	43.1	43.7
12.8	2.34	31.2	35.9	37.2	38.4	39.4	40.3	41.2	41.9	42.6	43.3	43.9
EFFICIENCY, η = 65%												
11.0	2.73	30.3	34.9	36.3	37.4	38.5	39.4	40.2	41.0	41.6	42.3	42.9
11.2	2.68	30.4	35.1	36.4	37.6	38.6	39.5	40.4	41.1	41.8	42.4	43.0
11.4	2.63	30.6	35.2	36.6	37.7	38.8	39.7	40.5	41.3	42.0	42.6	43.2
11.6	2.59	30.7	35.4	36.7	37.9	38.9	39.8	40.7	41.4	42.1	42.8	43.4
11.8	2.54	30.9	35.5	36.9	38.0	39.1	40.0	40.8	41.6	42.2	42.9	43.5
12.0	2.50	31.0	35.7	37.0	38.2	39.2	40.1	40.9	41.7	42.4	43.0	43.6
12.2	2.46	31.2	35.8	37.2	38.3	39.3	40.3	41.1	41.8	42.5	43.2	43.8
12.4	2.42	31.3	36.0	37.3	38.5	39.5	40.4	41.2	42.0	42.7	43.3	43.9
12.6	2.38	31.4	36.1	37.5	38.6	39.6	40.6	41.4	42.1	42.8	43.5	44.1
12.8	2.34	31.6	36.3	37.6	38.7	39.8	40.7	41.5	42.3	43.0	43.6	44.2

Frequency (GHz)	*Wavelength* (cm)	*Diameter* (cm)										
		35	60	70	80	90	100	110	120	130	140	150
EFFICIENCY, η = 70%												
11.0	2.73	30.6	35.3	36.6	37.8	38.8	39.7	40.5	41.3	42.0	42.6	43.2
11.2	2.68	30.7	35.4	36.8	37.9	38.9	39.8	40.7	41.4	42.1	42.8	43.4
11.4	2.63	30.9	35.6	36.9	38.1	39.1	40.0	40.8	41.6	42.3	42.9	43.5
11.6	2.59	31.0	35.7	37.1	38.2	39.2	40.2	41.0	41.7	42.4	43.1	43.7
11.8	2.54	31.2	35.9	37.2	38.4	39.4	40.3	41.1	41.9	42.6	43.2	43.8
12.0	2.50	31.3	36.0	37.4	38.5	39.5	40.4	41.3	42.0	42.7	43.4	44.0
12.2	2.46	31.5	36.2	37.5	38.7	39.7	40.6	41.4	42.2	42.9	43.5	44.1
12.4	2.42	31.6	36.3	37.6	38.8	39.8	40.7	41.6	42.3	43.0	43.7	44.3
12.6	2.38	31.8	36.4	37.8	38.9	40.0	40.9	41.7	42.5	43.2	43.8	44.4
12.8	2.34	31.9	36.6	37.9	39.1	40.1	41.0	41.8	42.6	43.3	43.9	44.5

EFFICIENCY, η = 75%												
11.0	2.73	30.9	35.6	36.9	38.1	39.1	40.0	40.8	41.6	42.3	42.9	43.5
11.2	2.68	31.0	35.7	37.1	38.2	39.2	40.1	41.0	41.7	42.4	43.1	43.7
11.4	2.63	31.2	35.9	37.2	38.4	39.4	40.3	41.1	41.9	42.6	43.2	43.8
11.6	2.59	31.3	36.0	37.4	38.5	39.5	40.5	41.3	42.0	42.7	43.4	44.0
11.8	2.54	31.5	36.2	37.5	38.7	39.7	40.6	41.4	42.2	42.9	43.5	44.1
12.0	2.50	31.6	36.3	37.7	38.8	39.8	40.7	41.6	42.3	43.0	43.7	44.3
12.2	2.46	31.8	36.5	37.8	39.0	40.0	40.9	41.7	42.5	43.2	43.8	44.4
12.4	2.42	31.9	36.6	37.9	39.1	40.1	41.0	41.9	42.6	43.3	44.0	44.6
12.6	2.38	32.1	36.7	38.1	39.2	40.3	41.2	42.0	42.8	43.5	44.1	44.7
12.8	2.34	32.2	36.9	38.2	39.4	40.4	41.3	42.1	42.9	43.6	44.2	44.8
EFFICIENCY, η = 80%												
11.0	2.73	31.2	35.8	37.2	38.3	39.4	40.3	41.1	41.9	42.6	43.2	43.8
11.2	2.68	31.3	36.0	37.3	38.5	39.5	40.4	41.3	42.0	42.7	43.4	44.0
11.4	2.63	31.5	36.1	37.5	38.6	39.7	40.6	41.4	42.2	42.9	43.5	44.1
11.6	2.59	31.6	36.3	37.6	38.8	39.8	40.7	41.6	42.3	43.0	43.7	44.3
11.8	2.54	31.8	36.4	37.8	38.9	40.0	40.9	41.7	42.5	43.2	43.8	44.4
12.0	2.50	31.9	36.6	37.9	39.1	40.1	41.0	41.9	42.6	43.3	44.0	44.6
12.2	2.46	32.1	36.7	38.1	39.2	40.3	41.2	42.0	42.8	43.5	44.1	44.7
12.4	2.42	32.2	36.9	38.2	39.4	40.4	41.3	42.1	42.9	43.6	44.2	44.8
12.6	2.38	32.3	37.0	38.4	39.5	40.5	41.5	42.3	43.0	43.7	44.4	45.0
12.8	2.34	32.5	37.2	38.5	39.7	40.7	41.6	42.4	43.2	43.9	44.5	45.1

Appendix 7

EFFECTIVE AREAS OF PARABOLIC ANTENNAS

The output of a receiving antenna (P_r) of course depends on the level of the input signal but also on the area and efficiency of the antenna iteself. Not all the signal on the antenna surface is successfully directed into the waveguide of the antenna output system so we account for this by use of a fraction generally designated by the Greek, η, so resulting in an area somewhat less than the physical area. Accordingly the *effective area* (A_{eff}) of an antenna is equal to ηA where A is the physical area.

Since:

$$A_{eff} = \eta A \quad \text{and} \quad A = \frac{\pi D^2}{4}$$

where D is the antenna diameter in metres, then in decibels,

$$A_{eff} = 10 \log \left(\frac{\pi D^2 \eta}{4} \right).$$

Table A7.1 quotes the values of A_{eff} for a practical range of antenna diameters and efficiencies.

Table A7.1
EFFECTIVE AREAS OF PARABOLIC ANTENNAS

Dish diameter (cm)	A_{eff} expressed in decibels						
	efficiency, η (%)						
	50	55	60	65	70	75	80
30	−14.5	−14.1	−13.7	−13.4	−13.1	−12.8	−12.5
40	−12.0	−11.6	−11.2	−10.9	−10.6	−10.3	−10.0
50	−10.1	−9.7	−9.3	−8.9	−8.6	−8.3	−8.0
60	−8.5	−8.1	−7.7	−7.4	−7.0	−6.7	−6.5
70	−7.2	−6.7	−6.4	−6.0	−5.7	−5.4	−5.1
80	−6.0	−5.6	−5.2	−4.9	−4.5	−4.2	−4.0
90	−5.0	−4.6	−4.2	−3.8	−3.5	−3.2	−2.9
100	−4.1	−3.6	−3.3	−2.9	−2.6	−2.3	−2.0
120	−2.5	−2.1	−1.7	−1.3	−1.0	−0.7	−0.4
140	−1.1	−0.7	−0.3	0	0.3	0.6	0.9
160	0	0.4	0.8	1.2	1.5	1.8	2.1
180	1.0	1.5	1.8	2.2	2.5	2.8	3.1

See also Section 4.4.

Appendix 8
ANTENNA POINTING

A8.1 Azimuth and Elevation

The principles of adjusting a dish to receive from a particular satellite are discussed in Section 4.8.1. This appendix presents four tables which together enable the azimuth and elevation angles for a dish to be determined for any site in Europe. The formulae from which the tables have been developed follow, examples of their use are given in Section 4.8.1.1.

Let the latitude of the receiving station = θ_r, the longitude = ϕ_r. Let the longitude of the satellite = ϕ_s (East is negative). Then longitude difference between satellite and receiving station:

$$\phi_d = \phi_s - \phi_r \tag{1}$$

Let r = radius of earth (6378 km); h = height of satellite (35 786 km) and $x = \cos^{-1}(\cos\theta_r, \cos\phi_d)$. Then azimuth angle:

$$A_z = \tan^{-1}\left(\frac{\tan\phi_d}{\sin\theta_r}\right) + 180 \text{ degrees} \tag{2}$$

Elevation angle:

$$E_l = \tan^{-1}\left(\cot x - \frac{r}{r+h} \cdot \operatorname{cosec} x\right)$$

$$= \tan^{-1}(\cot x - 0.1513 \operatorname{cosec} x)$$

or

$$\tan^{-1}\left(\frac{1}{\tan x} - \frac{0.1513}{\sin x}\right) \text{ degrees} \tag{3}$$

Both formulae therefore involve tangents and sines only (note that $\cos^{-1}$ and $\tan^{-1}$ are synonymous with arcos and arctan).

(Tables overleaf)

Table A8.1 – AZIMUTH (ϕ_d POSITIVE)

Lat, θ_r, dgs	LONGITUDE DIFFERENCE, ϕ_d, degrees														
	0	5	10	15	20	25	30	35	40	45	50	55	60	65	70
38	180	188.1	196.0	203.5	210.6	217.1	223.2	228.7	233.7	238.4	242.7	246.7	250.4	254.0	257.4
39	180	187.9	195.7	203.1	210.0	216.5	222.5	228.1	233.1	237.8	242.2	246.2	250.0	253.6	257.1
40	180	187.8	195.3	202.6	209.5	216.0	221.9	227.4	232.5	237.3	241.7	245.8	249.6	253.3	256.8
41	180	187.6	195.0	202.2	209.0	215.4	221.3	226.9	232.0	236.7	241.2	245.3	249.3	253.0	256.6
42	180	187.4	194.8	201.8	208.5	214.9	220.8	226.3	231.4	236.2	240.7	244.9	248.9	252.7	256.3
43	180	187.3	194.5	201.4	208.1	214.4	220.2	225.8	230.9	235.7	240.2	244.5	248.5	252.4	256.1
44	180	187.2	194.2	201.1	207.7	213.9	219.7	225.2	230.4	235.2	239.8	244.1	248.1	252.1	255.8
45	180	187.1	194.0	200.8	207.2	213.4	219.2	224.7	229.9	234.7	239.3	243.7	247.8	251.8	255.6
46	180	186.9	193.8	200.4	206.8	213.0	218.8	224.2	229.4	234.3	238.9	243.3	247.4	251.5	255.3
47	180	186.8	193.6	200.1	206.5	212.5	218.3	223.8	228.9	233.8	238.5	242.9	247.1	251.2	255.1
48	180	186.7	193.3	199.8	206.1	212.1	217.8	223.3	228.5	233.4	238.1	242.5	246.8	250.9	254.9
49	180	186.6	193.2	199.5	205.7	211.7	217.4	222.9	228.0	233.0	237.7	242.1	246.5	250.6	254.6
50	180	186.5	193.0	199.3	205.4	211.3	217.0	222.4	227.6	232.5	237.3	241.8	246.1	250.3	254.4
51	180	186.4	192.8	199.0	205.1	211.0	216.6	222.0	227.2	232.1	236.9	241.4	245.8	250.1	254.2
52	180	186.3	192.6	198.8	204.8	210.6	216.2	221.6	226.8	231.8	236.5	241.1	245.5	249.8	254.0
53	180	186.3	192.5	198.5	204.5	210.3	215.9	221.2	226.4	231.4	236.2	240.8	245.2	249.6	253.8
54	180	186.2	192.3	198.3	204.2	210.0	215.5	220.9	226.0	231.0	235.8	240.5	245.0	249.3	253.6
55	180	186.1	192.1	198.1	204.0	209.7	215.2	220.5	225.7	230.7	235.5	240.2	244.7	249.1	253.4
56	180	186.0	192.0	197.9	203.7	209.4	214.9	220.2	225.3	230.3	235.2	239.9	244.4	248.9	253.2
57	180	186.0	191.9	197.7	203.5	209.1	214.5	219.9	225.0	230.0	234.9	239.6	244.2	248.6	253.0
58	180	185.9	191.7	197.5	203.2	208.8	214.2	219.5	224.7	229.7	234.6	239.3	243.9	248.4	252.8

Table A8.2 – AZIMUTH (ϕ_d NEGATIVE)

Lat, θ_r, dgs	LONGITUDE DIFFERENCE, ϕ_d, degrees														
	−0	−5	−10	−15	−20	−25	−30	−35	−40	−45	−50	−55	−60	−65	−70
38	180	171.9	164.0	156.5	149.4	142.9	136.8	131.3	126.3	121.6	117.3	113.3	109.6	106.0	102.6
39	180	172.1	164.3	156.9	150.0	143.5	137.5	131.9	126.9	122.2	117.8	113.8	110.0	106.4	102.9
40	180	172.2	164.7	157.4	150.5	144.0	138.1	132.6	127.5	122.7	118.3	114.2	110.4	106.7	103.2
41	180	172.4	165.0	157.8	151.0	144.6	138.7	133.1	128.0	123.3	118.8	114.7	110.7	107.0	103.4
42	180	172.6	165.2	158.2	151.5	145.1	139.2	133.7	128.6	123.8	119.3	115.1	111.1	107.3	103.7
43	180	172.7	165.5	158.6	151.9	145.6	139.8	134.2	129.1	124.3	119.8	115.5	111.5	107.6	103.9
44	180	172.8	165.8	158.9	152.3	146.1	140.3	134.8	129.6	124.8	120.2	115.9	111.9	107.9	104.2
45	180	172.9	166.0	159.2	152.8	146.6	140.8	135.3	130.1	125.3	120.7	116.3	112.2	108.2	104.4
46	180	173.1	166.2	159.6	153.2	147.0	141.2	135.8	130.6	125.7	121.1	116.7	112.6	108.5	104.7
47	180	173.2	166.4	159.9	153.5	147.5	141.7	136.2	131.1	126.2	121.5	117.1	112.9	108.8	104.9
48	180	173.3	166.7	160.2	153.9	147.9	142.2	136.7	131.5	126.6	121.9	117.5	113.2	109.1	105.1
49	180	173.4	166.8	160.5	154.3	148.3	142.6	137.1	132.0	127.0	122.3	117.9	113.5	109.4	105.4
50	180	173.5	167.0	160.7	154.6	148.7	143.0	137.6	132.4	127.5	122.7	118.2	113.9	109.7	105.6
51	180	173.6	167.2	161.0	154.9	149.0	143.4	138.0	132.8	127.9	123.1	118.6	114.2	109.9	105.8
52	180	173.7	167.4	161.2	155.2	149.4	143.8	138.4	133.2	128.2	123.5	118.9	114.5	110.2	106.0
53	180	173.7	167.5	161.5	155.5	149.7	144.1	138.8	133.6	128.6	123.8	119.2	114.8	110.4	106.2
54	180	173.8	167.7	161.7	155.8	150.0	144.5	139.1	134.0	129.0	124.2	119.5	115.0	110.7	106.4
55	180	173.9	167.9	161.9	156.0	150.3	144.8	139.5	134.3	129.3	124.5	119.8	115.3	110.9	106.6
56	180	174.0	168.0	162.1	156.3	150.6	145.1	139.8	134.7	129.7	124.8	120.1	115.6	111.1	106.8
57	180	174.0	168.1	162.3	156.5	150.9	145.5	140.1	135.0	130.0	125.1	120.4	115.8	111.4	107.0
58	180	174.1	168.3	162.5	156.8	151.2	145.8	140.5	135.3	130.3	125.4	120.7	116.1	111.6	107.2

Table A8.3 – ELEVATION (ϕ_d POSITIVE)

Lat, θ_r, dgs	LONGITUDE DIFFERENCE, ϕ_d, degrees														
	0	5	10	15	20	25	30	35	40	45	50	55	60	65	70
38	46.0	45.7	44.7	43.2	41.2	38.8	36.0	32.9	29.6	26.1	22.4	18.6	14.8	10.9	7.0
39	44.8	44.5	43.7	42.2	40.3	37.9	35.2	32.2	28.9	25.5	21.9	18.2	14.4	10.6	6.8
40	43.7	43.4	42.6	41.2	39.3	37.0	34.4	31.5	28.3	24.9	21.4	17.8	14.1	10.3	6.5
41	42.6	42.3	41.5	40.2	38.4	36.1	33.6	30.7	27.6	24.3	20.9	17.3	13.7	10.0	6.3
42	41.5	41.2	40.4	39.1	37.4	35.2	32.7	30.0	26.9	23.7	20.4	16.9	13.3	9.7	6.1
43	40.4	40.1	39.4	38.1	36.4	34.3	31.9	29.2	26.3	23.1	19.9	16.5	13.0	9.4	5.8
44	39.3	39.0	38.3	37.1	35.4	33.4	31.1	28.5	25.6	22.5	19.3	16.0	12.6	9.1	5.6
45	38.2	37.9	37.2	36.1	34.5	32.5	30.3	27.7	24.9	21.9	18.8	15.5	12.2	8.8	5.3
46	37.1	36.8	36.1	35.0	33.5	31.6	29.4	26.9	24.2	21.3	18.3	15.1	11.8	8.5	5.1
47	36.0	35.7	35.1	34.0	32.5	30.7	28.6	26.2	23.5	20.7	17.7	14.6	11.4	8.1	4.8
48	34.9	34.7	34.0	33.0	31.6	29.8	27.7	25.4	22.8	20.1	17.2	14.1	11.0	7.8	4.6
49	33.8	33.6	33.0	31.9	30.6	28.9	26.9	24.6	22.1	19.4	16.6	13.7	10.6	7.5	4.3
50	32.7	32.5	31.9	30.9	29.6	28.0	26.0	23.8	21.4	18.8	16.0	13.2	10.2	7.1	4.0
51	31.6	31.4	30.8	29.9	28.6	27.0	25.2	23.0	20.7	18.2	15.5	12.7	9.8	6.8	3.7
52	30.5	30.3	29.8	28.9	27.6	26.1	24.3	22.2	20.0	17.5	14.9	12.2	9.3	6.4	3.5
53	29.4	29.3	28.7	27.9	26.7	25.2	23.4	21.4	19.2	16.9	14.3	11.7	8.9	6.1	3.2
54	28.3	28.2	27.7	26.8	25.7	24.3	22.6	20.6	18.5	16.2	13.7	11.2	8.5	5.7	2.9
55	27.3	27.1	26.6	25.8	24.7	23.3	21.7	19.8	17.8	15.5	13.2	10.7	8.1	5.4	2.6
56	26.2	26.0	25.6	24.8	23.7	22.4	20.8	19.0	17.0	14.9	12.6	10.1	7.6	5.0	2.3
57	25.1	25.0	24.5	23.8	22.8	21.5	20.0	18.2	16.3	14.2	12.0	9.6	7.2	4.6	2.0
58	24.1	23.9	23.5	22.8	21.8	20.6	19.1	17.4	15.6	13.5	11.4	9.1	6.7	4.3	1.7

Table A8.4 – ELEVATION (ϕ_d NEGATIVE)

Lat, θ_r, dgs	LONGITUDE DIFFERENCE, ϕ_d, degrees														
	−0	−5	−10	−15	−20	−25	−30	−35	−40	−45	−50	−55	−60	−65	−70
38	46.0	45.7	44.7	43.2	41.2	38.8	36.0	32.9	29.6	26.1	22.4	18.6	14.8	10.9	7.0
39	44.8	44.5	43.7	42.2	40.3	37.9	35.2	32.2	28.9	25.5	21.9	18.2	14.4	10.6	6.8
40	43.7	43.4	42.6	41.2	39.3	37.0	34.4	31.5	28.3	24.9	21.4	17.8	14.1	10.3	6.5
41	42.6	42.3	41.5	40.2	38.4	36.1	33.6	30.7	27.6	24.3	20.9	17.3	13.7	10.0	6.3
42	41.5	41.2	40.4	39.1	37.4	35.2	32.7	30.0	26.9	23.7	20.4	16.9	13.3	9.7	6.1
43	40.4	40.1	39.4	38.1	36.4	34.3	31.9	29.2	26.3	23.1	19.9	16.5	13.0	9.4	5.8
44	39.3	39.0	38.3	37.1	35.4	33.4	31.1	28.5	25.6	22.5	19.3	16.0	12.6	9.1	5.6
45	38.2	37.9	37.2	36.1	34.5	32.5	30.3	27.7	24.9	21.9	18.8	15.5	12.2	8.8	5.3
46	37.1	36.8	36.1	35.0	33.5	31.6	29.4	26.9	24.2	21.3	18.3	15.1	11.8	8.5	5.1
47	36.0	35.7	35.1	34.0	32.5	30.7	28.6	26.2	23.5	20.7	17.7	14.6	11.4	8.1	4.8
48	34.9	34.7	34.0	33.0	31.6	29.8	27.7	25.4	22.8	20.1	17.2	14.1	11.0	7.8	4.6
49	33.8	33.6	33.0	31.9	30.6	28.9	26.9	24.6	22.1	19.4	16.6	13.7	10.6	7.5	4.3
50	32.7	32.5	31.9	30.9	29.6	28.0	26.0	23.8	21.4	18.8	16.0	13.2	10.2	7.1	4.0
51	31.6	31.4	30.8	29.9	28.6	27.0	25.2	23.0	20.7	18.2	15.5	12.7	9.8	6.8	3.7
52	30.5	30.3	29.8	28.9	27.6	26.1	24.3	22.2	20.0	17.5	14.9	12.2	9.3	6.4	3.5
53	29.4	29.3	28.7	27.9	26.7	25.2	23.4	21.4	19.2	16.9	14.3	11.7	8.9	6.1	3.2
54	28.3	28.2	27.7	26.8	25.7	24.3	22.6	20.6	18.5	16.2	13.7	11.2	8.5	5.7	2.9
55	27.3	27.1	26.6	25.8	24.7	23.3	21.7	19.8	17.8	15.5	13.2	10.7	8.1	5.4	2.6
56	26.2	26.0	25.6	24.8	23.7	22.4	20.8	19.0	17.0	14.9	12.6	10.1	7.6	5.0	2.3
57	25.1	25.0	24.5	23.8	22.8	21.5	20.0	18.2	16.3	14.2	12.0	9.6	7.2	4.6	2.0
58	24.1	23.9	23.5	22.8	21.8	20.6	19.1	17.4	15.6	13.5	11.4	9.1	6.7	4.3	1.7

A8.2 Magnetic Variation (see also Sect.7.3.1.1)
The earth has two northern poles:

(i) the *geographic* or *true* North as indicated on any map, this is the point at the end of the axis on which the earth rotates;

(ii) the *geomagnetic* or *magnetic* North which is the point at which the earth's magnetic field cuts the earth's surface. The geomagnetic axis is at about 11 degrees to the geographic but unfortunately its position wanders slightly year by year.

The magnetic North pole produces a force which acts on a compass needle. On the other hand the true North pole has no physical attribute which we can use, hence when in setting up a satellite dish we need to know in which direction the geographic North lies, we can only find this via the magnetic North.

It is clear that the angular difference between the two Norths will vary according to the geographic location from which the measurement is being made. Table A8.5 therefore gives estimated magnetic variation figures for most of Western Europe, calculated for the year 1996. Because the figures are rounded to the nearest degree, they may be considered as being usable for a year or so before and beyond 1996 since the change is only about 10 minutes (0.167 degrees) per annum.

A8.3 Dish Declination Offset Angles (Sect.7.3.1.4)
The dish declination angle is illustrated in Figure 7.9 which shows that the sum of this angle and the elevation angle is equal to 90°. When a dish points due South then the difference in longitude, ϕ_d, between dish and pole of the arc is zero, hence from Section A8.1:

$$x = \cos^{-1}(\cos\theta \times \cos\phi_d)$$

but $$\phi_d = 0 \quad \therefore \cos\phi_d = 1$$

$$\therefore x = \cos^{-1}(\cos\theta) = \theta$$

and from Equation (3), elevation is equal to

$$\tan^{-1}\left(\frac{1}{\tan\theta} - \frac{0.1513}{\sin\theta}\right)$$

Table A8.5 – ESTIMATED MAGNETIC VARIATION (JANUARY 1996)

(Figures to be added to true bearing to obtain compass bearing)

LATITUDE, degrees	LONGITUDE, degrees ← W →											← E →				
	10	9	8	7	6	5	4	3	2	1	0	1	2	3	4	5
58					8	7	7	6	6							
57					8	7	6	6	5							
56					7	7	6	6	5							
55		9	8	8	7	6	6	5	5	4						
54	9	9	8	7	7	6	6	5	5	4	4					
53	9	8	8	7	7	6	5	5	4	4	4	3	3			
52		8	7	7	6	6	5	5	4	4	3	3	2			2
51						5	5	5	4	4	3	3	3	2	2	2
50					6	5				3	2	2	2	2	2	1
49									4	3	3	3	2	2	2	1
48						5	5	4	4	3	3	3	2	2	2	
47								4	3	3	3	2	2	2	2	
46										3	3	2	2	2	1	
45										3	2	2	2	2	1	
44										3	2	2	2	1	1	
43	6	6	5	5	5	4	4	3	3	3	2	2	2	1	1	
42		6	5	5	5	4	4	3	3	3	2	2	2	1	1	
41		6	5	5	4	4	4	3	3	3	2	2	2	1		
40		6	5	5	4	4	4	3	3	3	2					
39		6	5	5	4	4	4	3	3	2	2					
38		5	5	5	4	4	4	3	3	2						

$$= \tan^{-1} \frac{\cos\theta - 0.1513}{\sin\theta}$$

i.e. the elevation is related to a single variable, the latitude, θ.

If, using this formula, a graph is plotted of elevations over a range of latitudes for Europe (say, 36° – 60°) it is virtually a straight line conforming to the approximate relationship:

$$\text{Elevation} \approx 87.7 - 1.1 \times \text{latitude} \qquad (1)$$

$$\therefore \text{ Dish declination} \approx 90^\circ - \text{elevation} \approx 23 + 1.1 \times \text{latitude} \qquad (2)$$

Hence for any latitude a dish can be aligned by setting the elevation from Tables A8.3 – A8.4 or more easily by setting the declination from Equation (2).

To avoid calculation, what is most frequently used is a table of *dish declination offset angles.* These are angles which are added to the latitude to obtain the declination to which the dish is finally set. Table A8.6 gives the offset angles for European latitudes.

As an example, let us try out all the methods of calculating the setting-up angles for a polar-mount dish in, say, Brighton (UK) at approximately 51°N.

(i) From Table A8.3, elevation = 31.6°.

(ii) From Equation (1), elevation = 31.6°.

If the dish is calibrated for declination, then:

(iii) From Equation (2), declination = 58.4°.

(iv) From Table A8.6, declination = 51 + 7.5 = 58.5°,

a reasonably consistent set of results.

Table A8.6
DISH DECLINATION OFFSET ANGLES

Latitude (degrees)	Offset angle (degrees)
36	5.8
37	6.0
38	6.1
39	6.2
40	6.3
41	6.5
42	6.6
43	6.7
44	6.8
45	6.9
46	7.0
47	7.1
48	7.2
49	7.3
50	7.4
51	7.5
52	7.6
53	7.7
54	7.8
55	7.8
56	7.9
57	8.0
58	8.0
59	8.1
60	8.2

Appendix 9
PARABOLIC ANTENNA BEAMWIDTH

The theoretical width of the radiation beam of a transmitting parabolic antenna can be calculated. In this case however the formula is developed from a reasoning which involves Bessel functions, hence we must decline to get too deeply involved. At least we can use the formula to produce graphs of the beams of two practical antennas so that the effect of antenna size can be judged. Imagining the beam to be in the shape of a cone, let β degrees be half the cone apex angle, i.e. the angle with respect to the principal axis of the antenna aperture. The signal strengths at the edges of the cone can then be compared by:

$$E_\beta = J_1 \left[\frac{\pi d \sin \beta}{\lambda} \right] \times \frac{2\lambda \operatorname{cosec} \beta}{\pi d} \qquad (1)$$

where E_β is the signal strength relative to the maximum, β is the angle in degrees with respect to the principal axis of the antenna aperture, d is the antenna diameter (cm), λ is the transmission wavelength (cm) and J_1 is the first order Bessel function.

Note: the expression within the square brackets must first be evaluated, the Bessel function subsequently being obtained either mathematically or from published tables. Note also that when comparing two dishes of different sizes their maximum signal strengths are different, this is omitted from this calculation because we are only concerned with the effect of the angle β.

Consider two dishes, 60 cm and 120 cm diameter, both at 12 GHz for which λ = 2.5 cm. A typical calculation is as follows:

120 cm dish for $\beta = 0.6°$ (i.e. cone apex angle = 1.2°):

$$J_1 \left[\frac{\pi d \sin \beta}{\lambda} \right] = J_1 \left[\frac{\pi \times 120 \times \sin 0.6°}{2.5} \right] = J_1 [1.579] \; .$$

From Bessel Function Tables for J_1 we get 0.5595.

$$\therefore E_\beta = 0.5595 \times \frac{2 \times 2.5 \times \operatorname{cosec} 0.6°}{\pi \times 120} = 0.5595 \times 1.267 = 0.71$$

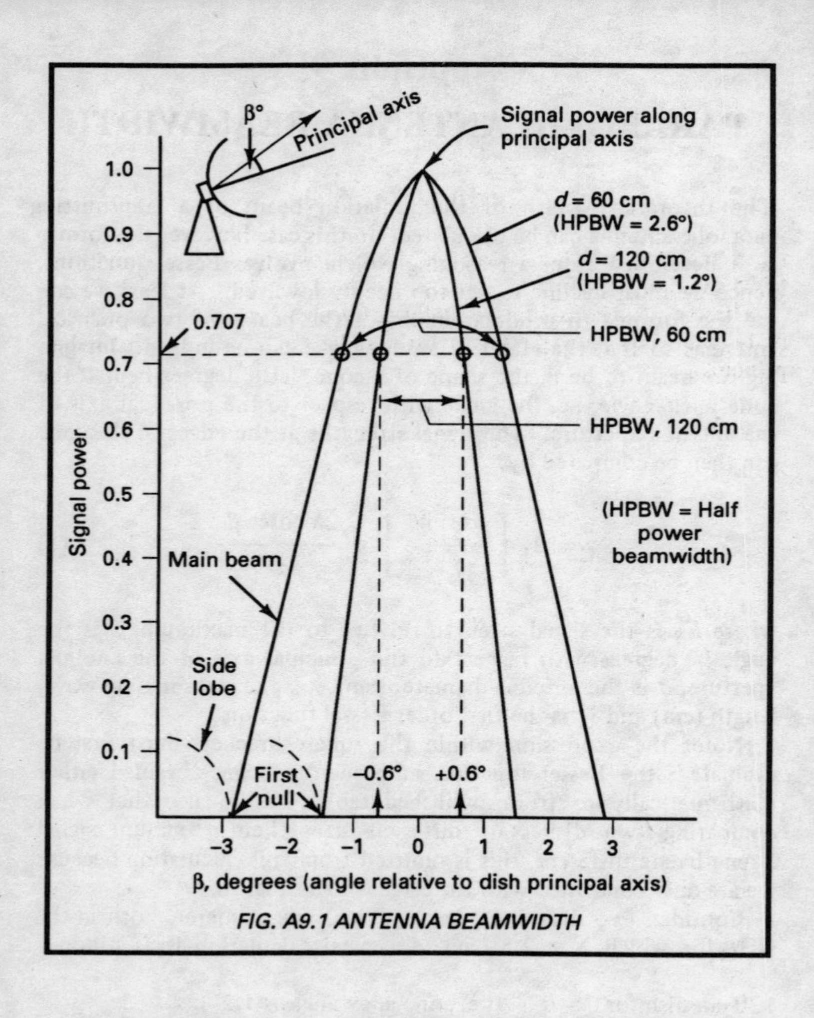

FIG. A9.1 ANTENNA BEAMWIDTH

Repeating this as required enables us to draw the two graphs shown in Figure A9.1. The effect of using an antenna of larger diameter is shown clearly in that its beam concentration is greater. By continuing the calculations beyond the value of β where the transmitted signal is zero (the first *null*), small *side lobes* appear on the graph but their levels are considerably lower than that of the main beam.

If we now consider the first null, i.e. the angle outside of which there is very little radiation, this must occur for any antenna when the Bessel function becomes zero, at which $J_1 = 3.83$, then

$$\frac{\pi d}{\lambda} \times \sin\beta = 3.83$$

$$\therefore \qquad \beta = \sin^{-1}\left(\frac{3.83 \times \lambda}{\pi d}\right) = \sin^{-1}\left(1.22 \times \frac{\lambda}{d}\right)$$

i.e. β increases as (λ/d) increases, hence for a narrow beamwidth, (d/λ) should be high.

These conclusions apply equally to receiving antennas for which it is important to receive as little radiation as possible off the principal axis.

Appendix 10
FREQUENCY MODULATION

Consider a carrier wave of frequency, f_c modulated by a sinusoidal wave of lower frequency, f_m with the equation, $v = V_m \sin \omega_m t$ where $\omega_m = 2\pi f_m$. Let the frequency deviation = Δf.

Since Δf varies according to the amplitude of the modulating wave, the deviation at any instant is $\Delta f \sin \omega_m t$

∴ Instantaneous frequency of f.m. wave (f) = (nominal frequency + deviation), i.e.

$$f = f_c + \Delta f \sin \omega_m t \qquad (1)$$

which gives the modulated wave frequency at any time, t.

Determining the *components* of an f.m. wave is a little complicated mathematically because the wave does not have a constant angular velocity. The outcome of doing so however is as follows.

Components of f.m. wave where V_c is the carrier amplitude and m the modulation index:

$$m = \frac{\Delta f}{f_m} \qquad (2)$$

(i) the carrier at an amplitude $V_c\left(1 - \frac{m^2}{4}\right)$

(ii) a pair of side frequencies ($f_c \pm f_m$), amplitude $mV_c/2$

(iii) a second pair of side frequencies ($f_c \pm 2f_m$), amplitude $m^2V_c/8$

(iv) higher order side frequencies, theoretically to infinity, ($f_c \pm 3f_m$), ($f_c \pm 4f_m$), etc.,

but reducing rapidly in amplitude and therefore can usually be neglected. This is a pointer to the fact that f.m. requires a large bandwidth which is generally of the order of 2 (peak frequency deviation + modulating frequency), i.e.

$$\text{bandwidth} \approx 2(\Delta f_{max} + f_m)\ \text{Hz} \qquad (3)$$

As mentioned in the main text, the maximum frequency deviation is set in the design stage. For an f.m. radio broadcast transmission a value of 75 kH is frequently quoted. If the maximum modulating frequency is 15 kHz then

$$\text{bandwidth} \approx 2(\Delta f_{\text{max}} + f_{\text{m}}) = 2(75 + 15) = 180\text{ kHz.}$$

For television a much greater value of Δf_{max} is required, say 6.5 MHz (13 MHz peak-peak) then for a maximum modulating frequency of, say 7 MHz

$$\text{bandwidth} \approx 2(6.5 + 7) = 27\text{ MHz.}$$

(Note that this is a theoretical consideration only. In practice other values of deviation may be used with the bandwidth restricted to 27 MHz. The actual values for any particular system are a compromise between signal power, overall quality, noise and bandwidth, hence Equation (3) can only be taken as a guide.)

Appendix 11

QUANTIZATION NOISE

Converting an analogue signal to digital involves quantizing into a finite number of fixed levels each separated from its neighbour by a step of amplitude q. The process cannot be reversed because, irrespective of the magnitude of q, each analogue signal amplitude value, when quantized results in a quantization level with an uncertainty of + or $-\ q/2$. In practice this gives rise to a *quantization noise* added to all quantized signals but although it can be reduced (see Sect. 5.5.4), it cannot be eliminated entirely.

Let us consider a system in which all quantization steps are the same and again each is equal to q. It is reasonable to expect that the amplitude of the signal being sampled may lie at any point within a quantization step, i.e. the quantizing error within a step may be considered to be a uniformly distributed random variable. Then

$$\text{the mean square error, } \bar{\epsilon}^2 = \frac{1}{q} \int_{-(q/2)}^{+(q/2)} \epsilon^2 \, d\epsilon$$

$$= \frac{1}{q} \left[\frac{\epsilon^2}{3} \right]_{-(q/2)}^{+(q/2)}$$

$$\text{which finally results in } \frac{1}{3q} \times \frac{q^3}{4} = \frac{q^2}{12}$$

The r.m.s. error is the square root of this, i.e. $q/\sqrt{12}$.

The noise power over one step is equal to the total power over all steps (P) divided by the number of levels (N), hence

$$\text{the } \textit{signal-to-quantization noise ratio} \text{ (SQNR)} = \frac{P^2}{(q^2\ /12)}$$

and since

$$P = q \times N,$$

$$\text{SQNR} = \frac{q^2 N^2 \times 12}{q^2} = 12N^2$$

or in decibels:

$$\text{SQNR} = 10.79 + 20 \log N.$$

Appendix 12
POLARIZATION OFFSET

Like the phenomenon from which it springs, *polarization offset* is not something which rapidly becomes crystal clear, especially when we find the term *skew* is also involved. In fact getting to grips with the concept is one of our more difficult undertakings. However polarization offset figures are frequently to be found with those for azimuth and elevation because they do affect the LNB or the dish as a whole. It is therefore desirable that we should have some knowledge of how they arise and are used.

The *propagation plane* of a wave is that which contains both the direction of propagation and the direction of electric vibrations. When facing an incoming vertically polarized wave therefore, the polarization plane might be imagined as a sheet of paper held vertically and end-on. A polarization angle arises when the polarization planes of satellite and receiving dish are not in line and this occurs for all receiving installation positions not at the same longitude as the satellite. The angular difference is known as the *polarization angle.* By rotating the receiving dish or LNB the polarization planes may be brought into line. The amount of rotation needed is called the *polarization offset* or *skew*. The mysterious "skew" simply indicates a divergence from the required direction or position (askew). Figures for this are given in Table A12.1, they are to the nearest degree and apply whether ϕ_d is positive or negative.

The formula for calculation of polarization angle and therefore the polarization offset required to counteract the effect is more than a little complicated. However with insignificant loss of accuracy, it can be reduced to a simple one:

$$\text{polarization offset} = \tan^{-1}\left\{\frac{\sin \phi_d}{\tan \theta_r}\right\} \text{ degrees},$$

where ϕ_d is the longitudinal difference between satellite and receiving locations (Sect.4.8) and θ_r is the latitude of the receiving location.

To use the table, first calculate ϕ_d, look up the polarization offset value, then standing behind the dish and looking towards the satellite:

Table A12.1 – POLARIZATION OFFSET (degrees)

Lat, θ_r, dgs	LONGITUDE DIFFERENCE, ϕ_d, degrees														
	0	5	10	15	20	25	30	35	40	45	50	55	60	65	70
38	0	6	13	18	24	28	33	36	39	42	44	46	48	49	50
39	0	6	12	18	23	28	32	35	38	41	43	45	47	48	49
40	0	6	12	17	22	27	31	34	38	40	42	44	46	47	48
41	0	6	11	17	22	26	30	33	37	39	41	43	45	46	47
42	0	6	11	16	21	25	29	33	36	38	40	42	44	45	46
43	0	5	11	16	20	24	28	32	35	37	39	41	43	44	45
44	0	5	10	15	20	24	27	31	34	36	38	40	42	43	44
45	0	5	10	15	19	23	27	30	33	35	38	39	41	42	43
46	0	5	10	14	18	22	26	29	32	34	37	38	40	41	42
47	0	5	9	14	18	22	25	28	31	33	36	37	39	40	41
48	0	5	9	13	17	21	24	27	30	33	35	36	38	39	40
49	0	4	9	13	17	20	24	27	29	32	34	36	37	38	39
50	0	4	8	12	16	20	23	26	28	31	33	35	36	37	38
51	0	4	8	12	16	19	22	25	28	30	32	34	35	36	37
52	0	4	8	11	15	18	21	24	27	29	31	33	34	35	36
53	0	4	8	11	15	18	21	23	26	28	30	32	33	34	35
54	0	4	7	11	14	17	20	23	25	27	29	31	32	33	34
55	0	4	7	10	14	17	19	22	24	26	28	30	31	32	33
56	0	3	7	10	13	16	19	21	23	26	27	29	30	31	32
57	0	3	6	10	13	15	18	20	23	25	26	28	29	31	31
58	0	3	6	9	12	15	17	20	22	24	26	27	28	30	30

rotation should be clockwise if the satellite lies to the West

(ϕ_d positive)

rotation should be anticlockwise if the satellite lies to the East

(ϕ_d negative)

In practice where there is remote control of the LNB for changing the polarization there will also be a skew control for fine tuning. This control usually sorts out all the polarization discrepancies together including the effects of depolarization on linear waves mentioned in Section 4.5. The polarization offset is therefore automatically taken into account.

Index

(See also Appendix 1 – Glossary of Satellite Communication Terms)